ACTIVE LEARNING FOR DIGITAL TRANSFORMATION IN HEALTHCARE EDUCATION, TRAINING, AND RESEARCH

ACTIVE LEARNING FOR DIGITAL TRANSFORMATION IN HEALTHCARE EDUCATION, TRAINING, AND RESEARCH

Edited by

MILTIADIS D. LYTRAS
College of Engineering, Effat University, Jeddah, Saudi Arabia

CRISTINA VAZ DE ALMEIDA
Instituto Superior de Ciências Sociais e Políticas (ISCSP), Lisbon, Portugal

Portuguese Health Literacy Society (SPLS), Lisbon, Portugal

ELSEVIER

ACADEMIC PRESS
An imprint of Elsevier

Academic Press is an imprint of Elsevier
125 London Wall, London EC2Y 5AS, United Kingdom
525 B Street, Suite 1650, San Diego, CA 92101, United States
50 Hampshire Street, 5th Floor, Cambridge, MA 02139, United States
The Boulevard, Langford Lane, Kidlington, Oxford OX5 1GB, United Kingdom

Notices
Knowledge and best practice in this field are constantly changing. As new research and experience
broaden our understanding, changes in research methods, professional practices, or medical treatment
may become necessary.

Practitioners and researchers must always rely on their own experience and knowledge in evaluating and using
any information, methods, compounds, or experiments described herein. In using such information or methods
they should be mindful of their own safety and the safety of others, including parties for whom they have a
professional responsibility.

To the fullest extent of the law, neither the Publisher nor the authors, contributors, or editors, assume any
liability for any injury and/or damage to persons or property as a matter of products liability, negligence or otherwise,
or from any use or operation of any methods, products, instructions, or ideas contained in the material herein.

ISBN 978-0-443-15248-1

For information on all Academic Press publications
visit our website at https://www.elsevier.com/books-and-journals

Publisher: Stacy Masucci
Acquisitions Editor: Linda Versteeg-Buschman
Editorial Project Manager: Sara Pianavilla
Production Project Manager: Sajana Devasi P K
Cover Designer: Greg Harris

Typeset by STRAIVE, India

Working together
to grow libraries in
developing countries

www.elsevier.com • www.bookaid.org

Contents

Contributors

Sérgio Filipe Silva Abrunheiro
The Coimbra Hospital and University Centre, Coimbra, Portugal

Basim S. Alsaywid
Saudi National Institute of Health, Riyadh, Saudi Arabia

Berta Maria Jesus Augusto
The Coimbra Hospital and University Centre, Coimbra, Portugal

Marta Barroca
Group of Health Centers (ACES) of the Tagus Estuary, Administration Regional Health (ARS) of Lisbon and Tagus Valley; Faculty of Medicine of Lisbon; Health Literacy from Ispa—University Institute

Manuel José Damásio
Centre for Research in Applied Communication Culture and New Technologies, Lusófona University, Lisbon, Portugal

Pedro Miguel de Almeida Melo
Universidade Católica Portuguesa, Centre for Interdisciplinary Research in Health/Institute of Health Sciences, Porto, Portugal

Rita Espanha
Communication, Culture and Information Technologies, Iscte-University Institute of Lisbon, Lisbon, Portugal

Sandra Laia Esteves
Regional Health Administration of Lisbon and Tagus Valley; Fiscal Council of the Portuguese Society of Health Literacy; Health Literacy from ISPA; Management in Health

Carlos Manuel Santos Fernandes
The Coimbra Hospital and University Centre, Coimbra, Portugal

Rui Brito Fonseca
The School of Education and Human Development, ISEC Lisboa, Lisbon, Portugal

Diogo Franco Santos
General and Family Medicine Specialty Training; Medicine at NOVA Medical School; Health Literacy

Francisco Garcia
Communication Sciences, Iscte-University Institute of Lisbon, Lisbon, Portugal

Sara Henriques
Centre for Research in Applied Communication Culture and New Technologies, Lusófona University, Lisbon, Portugal

Abdulrahman Housawi
Saudi Commission for the Health Specialties, Riyadh, Saudi Arabia

Helena Lima
University of Porto, Porto, Portugal

Miltiadis D. Lytras
College of Engineering, Effat University, Jeddah, Saudi Arabia

Patricia Martins
Portuguese Association for the Promotion of Public Health (APPSP); Regional Administration of Health of Lisbon and Tagus Valley (ARSLVT) to exercise functions in the Health Unit Public Arnaldo Sampaio of the Group of Health Centers (ACES) Arco Ribeirinho

Helena Águeda Marujo
Instituto Superior de Ciências Sociais e Politicas—ISCSP-CAPP, Center for Administration and Public Policies—CAPP, UNESCO Chair on Education for Global Peace Sustainability, University of Lisbon, Lisbon, Portugal

Nour Mheidly
Department of Communication, University of Illinois Chicago, Chicago, IL, United States

Tânia Manuel Moço Morgado
Centro Hospitalar e Universitário de Coimbra; Health Sciences Research Unit: Nursing (UICISA: E), Nursing School of Coimbra, Coimbra; School of Health Sciences, Polytechnic of Leiria, Leiria; Center for Health Technology and Services Research (CINTESIS), Porto Nursing School (ESEP), Porto, Portugal

Hugo Leiria Neves
Health Sciences Research Unit: Nursing (UICISA: E), Nursing School of Coimbra; Portugal Centre for Evidence Based Practice: A Joanna Briggs Institute Centre of Excellence, Coimbra, Portugal

Cecilia Nunes
SPLS—Portuguese Society of Health Literacy, Lisbon, Portugal

Hernâni Zão Oliveira
Department of Management and Innovation, Colégio do Espírito Santo, University of Évora, Évora, Portugal

Nuno Patraquim
University of Porto, Porto, Portugal

Joana Sofia Dias Pereira Sousa
School of Health Sciences, Polytechnic of Leiria; Center for Innovative Care and Health Technology - CiTechCare, Leiria, Portugal

Diana Pinheiro
SPLS—Portuguese Society of Health Literacy, Lisbon, Portugal; WFCMS—World Federation of Chinese Medicine Societies, Beijing, China

Manuela Soares Rodrigues
SPLS—Portuguese Society of Health Literacy, Lisbon; Egas Moniz – Higher Education Cooperative, CRL, Almada, Portugal

Patrícia Rodrigues
SPLS—Portuguese Society of Health Literacy, Lisbon, Portugal

Pedro Joel Rosa
Lusófona University, HEI-Lab: Digital Human–Environment Interaction Labs, Lisbon; Instituto Superior Manuel Teixeira Gomes (ISMAT), Portimão, Portugal

Rosa Carla Gomes Silva
Center for Health Technology and Services Research (CINTESIS), Porto Nursing School (ESEP), Porto; Portugal Centre for Evidence Based Practice: A Joanna Briggs Institute Centre of Excellence, Coimbra, Portugal

Cristina Vaz de Almeida
Instituto Superior de Ciências Sociais e Políticas (ISCSP); Portuguese Health Literacy Society (SPLS), Lisbon, Portugal

Ana Veiga
Fiscal Council of the Portuguese Society of Health Literacy; Dr. Gama Institute of Ophthalmology; Escola Superior de Enfermagem de Lisboa (ESEL)

Filipa Isabel Quaresma Santos Ventura
Health Sciences Research Unit: Nursing (UICISA: E); Nursing School of Coimbra, Coimbra, Portugal

CHAPTER 1

Active learning in healthcare education, training, and research: A digital transformation primer

Miltiadis D. Lytras[a] and Abdulrahman Housawi[b]
[a]College of Engineering, Effat University, Jeddah, Saudi Arabia
[b]Saudi Commission for the Health Specialties, Riyadh, Saudi Arabia

1 Foundations of active learning

Active learning is a holistic philosophy for the integration of student-centric, personalized, and objectives-oriented exploration of learning content within a given and enriched learning environment where individual and team learning paths enhance problem solving capability. This is our personal definition of the phenomenon and this high-level abstraction can be detailed further. In Fig. 1 below, we summarize our philosophical logical outlet and definition of active learning. Our key idea is that active learning defines a synergetic, collaborative, and exploratory learning design where several core components contribute to the learning outcome and justify the impact of education (Lytras, Papadopoulou, & Sarirete, 2020; Misseyanni, Marouli, Papadopoulou, Lytras, & Gastardo, 2016; Misseyanni, Papadopoulou, Marouli, & Lytras, 2018; Naeve, Yli-Luoma, Kravcik, & Lytras, 2008; Sairete et al., 2021; Spruit & Lytras, 2018).

In the next section we elaborate on the key dimensions of the active learning philosophy in higher education.

Student centric: Active learning is student centric. It is envisioning the coverage of the learning needs of students that are tailored to well-defined components of their learning profiles. One of the greatest challenges in higher education in our days is the systematic, monolithic, "one-size-fits-all" learning approach and strategy. Students with different learning needs, skills, competencies, and talents are struggling to be role-players in a curriculum-oriented training strategy with a unified, uniform learning delivery strategy. In my opinion, a new era of student-centric active learning strategy needs to calibrate the characteristics of each and every student. In this direction, a systematic profiling of students' characteristics including needs and desired outcomes has to be supported also with technology services. Learning analytics solutions, for example, can address meaningful learning paths for individuals. Additionally, the typical classroom-based approach to training must be critically reconsidered. The focus on single content modules without

1

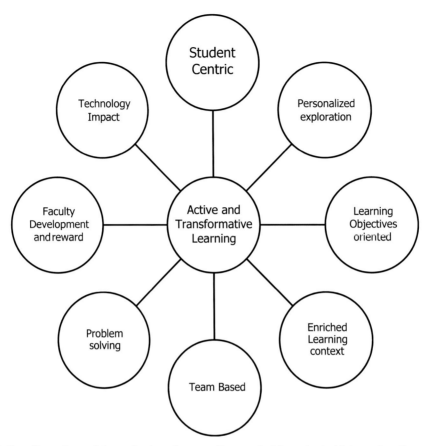

Fig. 1 Key dimensions of the active learning paradigm and philosophy in higher education.

exploratory journeys set barriers and limitations to the learning potential and the learning footprint of education.

Personalized exploration: Higher education is characterized by a myopic half-done learning strategy. For the last 10 years, at least, the development of educational and learning curricula was deterministic and was organized around learning outcomes and objectives. Competencies and skills were initially ignored and later included as a wishful thinking perspective. At the same moment, a huge gap in terms of industry requirements for professionals and graduates to the knowledge communicated in higher education was, on purpose, undervalued. This resulted in a really pragmatic situation in which the gap between academia and industry is challenging the development of a huge parallel to higher education, marketplace of certification of industry skills.

Learning objectives oriented: Programs in higher education are currently characterized by an emphasis on partial learning objectives. The underlying theories for this direction are used as a context, but the problem remains that learning objectives can be hardly measured or objectively reflected in learning skills and competencies. The body of knowledge and the taxonomy of learning objectives synthesize in an incomplete manner in many training programs without an overarching strategy for the overall impact of learning. This is not a simple philosophical discussion, but at the same moment, it serves also as a triggering turnaround point for the next generation of higher education. In our humble opinion, in recent years, the higher education administration has slowly responded to the critical requirements and the new facts of the "service" component of higher education. One of the most remarkable that are continuously underestimated is the capacity of learners in our time to compare in multidimensional ways the quality, the characteristics, and the added value of training programs.

Enriched learning context: The learning context in higher education has been multidimensionally enriched in the last few years. The experience of COVID-19 and the shift of learning mode to a distant form brought into reality several pulp fiction scenarios for technology-enhanced delivery of training and learning. Furthermore, the recent arrival of new technologies including virtual and augmented reality as well as the metaverse provides new opportunities for a more sophisticated learning context. The assumptions about time and place have also significantly enriched the learning context. New ideas about flexible learning and certification of competencies and skills from massive open online courses platforms or industrial players' approaches put significant pressure on the traditional higher education institutions. Novel ideas about the dynamic composition of the learning context are a key challenge for the new generational active learning in healthcare education. A new skills-oriented, team-based, open, and collaborative learning context needs to be established.

Team-based: The new active-learning paradigm in healthcare education needs to build strong anchors and pillars on team-based education. The new job profiles and the significant job requirements from the industry set role-playing team capabilities for young students. The higher education institutions must prepare the new skills force for this reality and bold developments shall transform rigid procedures and authoritarian higher education in new directions. Some examples include:

- Team-based assessments in health education,
- Team and role-playing evaluations,
- The development of competence-based educational curricula for individuals and teams, and
- Deployment of technology-enhanced learning for the design and implementation of collaborative active learning spaces.

Problem-solving and decision-making enhancement: The problem-solving capability is a key target of active learning in higher education. The design of educational programs and learning experience that will lead to the development of strong, resilient, and sustainable problem-solving capabilities must be a key priority of the active learning strategy in higher education institutions. The evolution of a new era of critical thinking skills building in association with the enhancement of decision-making capability must be set in the top priorities of new educational strategies in higher education institutions. From a practical point of view, this strategic objective should be translated into new training programs with significant know-how transfer and association with the practices and standards of the industry.

Faculty development, recognition, and reward: The transition to an effective active learning strategy in higher education institutions requires a bold faculty development, recognition, and reward strategy. It must be understood by the stakeholders and the administration that the empowerment of faculty with new skills and competencies, the promotion of a research-based culture, and the exploitation of a knowledge translation strategy by the faculty must be integrated with a robust strategy of recognition and reward. Active learning philosophy is not a mechanical approach. The contribution of the human factor is a catalyst for the potential impact and the multiplier of the added value. Designing fair faculty performance review procedures with significant rewards and resources for the faculty who applied active learning is a key action. Recognizing the critical contribution of the faculty to the strategic objectives of the higher education administration and involving the faculty in key decisions is another aspect of the active learning paradigm. The responsibility for the implementation of active learning must be shared and equally supported with resources and rewards. We do need a new era of academic administration that will vision the pathways for the successful implementation of the new strategies and will apply a codesign and coresponsibility mentality in higher education institutions.

Technology impact: The mechanics of active learning require also a bold technological supporting component. The variety of technology-enhanced learning tools, and the new highly promising value propositions of artificial intelligence in education, learning analytics, metaverse, open knowledge, free and open-source educational tools, cloud-based learning systems, educational repositories, etc., are promoting the ideas and engagement modes of active learning. Special emphasis in higher education institutions must be also placed on the establishment and full financing of teaching and learning centers and technology-enhanced learning facilities, with experts, tools, and procedures capable of promoting the active learning strategy to its full potential.

2 A roadmap for implementation of active learning strategy in higher education

The adoption of active learning in higher education has been many times perceived as a wishful thinking scenario, an unpainful implementation without a systematic strategy and implementation plan.

The industrialization of higher education is an additional fact that poses additional pressures to the promotion of the active learning paradigm. Higher education institutions with standardized processes, rigid bureaucratic management schemas, and slow-moving reengineering of educational curricula seem to be well-tuned input–process–output systems without many times considering the needs and the requirements of the current times (Lytras et al., 2020; Lytras, Sarirete, & Stasinopoulos, 2020; Lytras, Serban, Ruiz, Ntanos, & Sarirete, 2022).

In this section, we provide a preliminary approach to a strategic roadmap for the implementation of an active learning strategy in higher education institutions (see Fig. 2). This approach is also updated with research-based evidence that will be shortly published.

2.1 Active learning strategy formulation/design

The effective adoption of active learning strategy in higher education institutions requires a bold institutional strategy. The stewardship of active learning within the organization requires an orchestration of diverse resources and roles. In our recommended approach, we propose five critical components, namely:

- Strategic Objectives Formulation
- Executive Committee Formulation
- Action Plan and Roadmap
- Maturity Assessment
- Use Cases/Areas of Improvement

The **Formulation of Strategic Objectives** must communicate in an institution-wide mode the priorities and the measurable objectives of the active learning strategy. It is true that in this area there is a critical need for the development of a novel taxonomy of strategic objectives in close alignment with the new dynamic role of higher education in modern society.

The **Executive Committee** is an institutional standing committee supervising the design and implementation of the active learning strategy. It is also responsible for the

Fig. 2 A proposed roadmap for implementation of active learning strategy in higher education.

execution of an Active Learning Maturity Assessment that will lead to the drafting of a strategic Road Map and Action Plan and the identification of a variety of use cases and areas of improvement related to the utilization of active learning in the institution.

Action Plan and Roadmap: The accountable academic leadership and administration together with the institutional bodies responsible for the learning strategy should jointly confirm an action plan and a roadmap of strategic milestones and desired developments. This can serve as an agreement for the utilization of resources within given time frames toward the implementation of the strategy. The monitoring of performance should also be a continuous process and priority, with full engagement of faculty.

Maturity Assessment: Active learning must be developed on the grounds of a systematic maturity for all the learning-related initiatives, developments, and engagements over time. The evolution of academic institutions over time has proven isolated actions and exemplary initiatives. The maturity assessment is targeted at the understanding of the current performance of activities related to instructional design, technology-enhanced learning, assessment, learning outcomes, personalized learning, exploratory active discovery of learning content, and competence-based education.

Use Cases/Areas of Improvement: As with any strategic initiative, active learning implementation needs a systematic and detailed justification of use cases of active learning activities and initiatives that will address areas of improvement and real challenges in the academic context.

2.2 Active learning strategy implementation

The implementation of the Road Map and the Action Plan for the adoption of the active learning strategy entails a variety of complementary, meaningful initiatives including:

- *Redesign of Educational Programs and Curricula*: The maturity assessment of active learning needs to be accompanied by the redesign and launch of updated and novel educational programs and curricula. This will reflect the overarching principles of the active learning paradigm and will also motivate students and faculty engagement to the new value proposition of higher education
- *Faculty Development Program:* The faculty must be considered the most critical strategic asset for the success of the active learning strategy. The current overload of faculty in higher education institutions and the variety of obligations in terms of scholarship, teaching, and service will require additional resources and significant institutional support. In this direction, faculty and academic administration should be coresponsible for the active learning strategy codesign.
- *Institutional Support and Resources:* The support of active learning with resources and institutional commitment is a key requirement. Examples of resources include but are not limited to reduction of course load for faculty, enhancement of the services and the infrastructure of the active learning center in the institution, funding for active learning-related research, and financing for active learning training resources.

- *Technology-Enhanced Learning*: The investment in TEL tools, services, and applications must be strategized also for the promotion of the active learning strategy. This will require a systematic integration of technology-enhanced learning tools within the higher education institution and the integration of all the relevant tools and workflows in a unique center. Furthermore, the availability of diverse technologies and TEL resources will also require a strategic plan for the utilization of TEL for active learning.

2.3 Active learning strategy impact

Active learning strategy should be also measurable in terms of impact and added value. From this point of view, the deployment of complementary methodological tools and services for the analysis of the impact of the active learning strategy is necessary. The following methods and approaches are indicative:

- Learning Analytics: The latest developments in learning analytics and the adoption of educational data mining and AI over significant educational data sets can be of very useful service for active learning impact measurement. Analyzing students' dropout rates, customizing learning content delivery on the basis of active learning strategies, the dynamic composition of competence-based education scenarios, and the dynamic exploration of active learning content based on personal preferences are only a few examples of the application of learning analytics. Designing also dynamic learning analytics dashboards for students, faculty, and higher education administrators is another bold initiative for measuring the impact of active learning.
- Key Performance Indicators and Benchmarks: The design of an institution-wide taxonomy of active learning key performance indicators (KPIs) and benchmarks will also allow the implementation of an accountable active learning strategy with distinct roles and action plans that can be measured and analyzed.
- Know-How Transfer: One of the greatest challenges for any higher education institution is the effective know-how transfer within, across, and beyond the organization. The establishment of knowledge management and dissemination strategies that will communicate the benchmarks and the best practices from the internal and external environment of the organization must be a systematic developmental process.
- Certification and Accreditation: The active learning philosophy must be also functional, interoperable, and ubiquitous. It requires also certification and accreditation. The embodiment of active learning strategies within a college or university or a training organization must exploit skills and competencies that are certified and also standards and procedures that can be accredited and enhanced.
- Entrepreneurship: Active learning should also focus on the diverse audiences within the academic environment and also should fulfill students' needs that are oriented to entrepreneurship and disruptive new businesses. From this point of view, the active learning strategy should promote critical thinking, problem solving, and robust

knowledge transfer from academia to the industry. In close relevance to the previous pillar, the active learning strategy

- Innovation and Sustainability: The promotion of bold strategic objectives including socially responsive sustainable economic development and innovation should be also seen as strategic footprints of an effective active learning strategy. Higher education institutions need to integrate the active learning approach into programs that promote creative thinking and lead to innovation, as well as should integrate technological developments as a core component of technology-driven value proposition for new job professions and skills required by the 2030 industry market and beyond 2050.
- Academia-Industry Observatory: The orchestration of academic strategies with industry priorities needs also to be considered as bold actions for active learning that promotes employability and contribution to society with a positive footprint. From this point of view, an observatory that will allow the dynamic sharing of opinions and priorities from the industry related to jobs, skills, and competencies is a priority.

3 Key challenges for healthcare education

The previous discussion for the active learning paradigm can be also customized to the context and the challenges of healthcare education. In this brief section, we provide a brief analysis of the current challenges in the context of healthcare education with a focus on a few high-level aspects that need to be addressed by the active learning strategy. A summary of our ideas is also presented in Fig. 3 below.

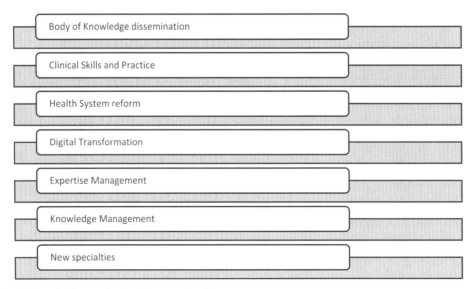

Fig. 3 Key challenges for healthcare education.

Body of Knowledge Dissemination: The body of knowledge on the health specialties is significant and with regular updates. This challenges the active learning strategy implementation. It requires diverse approaches for different granularities and specificities of the content that must be disseminated and understood by students. The variety and the details set also opportunities for different active learning scenarios and practices with a given strategic framework.

Clinical Skills and Practice: The other delicate component in health education is related to clinical skills development and clinical practice development. The availability of new technologies and the provision of sophisticated simulation tools, metaverse applications, augmented reality solutions, artificial intelligence agents, and knowledge management tools, among others, presents challenges for the strategic deployment of technologies for the enhancement of active learning toward clinical skills developments and clinical practice.

Health System Reform: The reform of the health system with the integration of new strategic objectives and new strategic pillars must be also updated in the educational curricula and in the educational and learning strategy. The rhythm of change and the pace of new developments should be integrated in a smooth way into educational programs.

Digital Transformation: The digital transformation in healthcare is promoting a unique value proposition for a new updated skillset for healthcare practitioners (Alsanea, 2012; Belliger & Krieger, 2018; Cresswell et al., 2013; Deloitte, 2018; Gopal, Suter-Crazzolara, Toldo, & Eberhardt, 2019; Haggerty, 2017; Kraus, Schiavone, Pluzhnikova, & Invernizzi, 2021). In this context, the active learning strategy should also append its strategic orientation to these new directions.

Expertise Management: One of the greatest challenges for health professions education is expertise management and know-how transfer. With the evolution of the internet and the web, the plethora of educational resources have made access to knowledge and training material much easier. In the same direction, knowledge networking and the matching of knowledge providers and knowledge consumers set the basis for a new era of sophisticated expertise management. Furthermore, the dynamic allocation of experts and training modules on demand is one additional challenge that can be facilitated by new technologies.

Knowledge Management: The knowledge intensity in healthcare and knowledge-driven decision making promote knowledge management as a significant and critical pillar of efficiency within the new healthcare strategic orientation. Academic institutions must realize the core contribution of knowledge management in the design, implementation, and provision of personalized educational curricula with active learning orientation.

New Specialties: Last but not least, the evolution of new technologies and the deep integration of machine learning, artificial intelligence and augmented reality, internet of things, cloud computing, etc., into the healthcare practice and education will develop new healthcare professions with a significant technological knowledge core. In this direction, healthcare academic institutions have to respond faster with new degrees and programs that reflect this new era.

4 Conclusions

Active learning in healthcare education, training, and research requires a robust strategy, codesigned by all the core stakeholders, and promoting the strategic objectives that reflect a sustainable educational approach. In this introductory chapter, we tried to highlight and to summarize thoughts and ideas that need further reflection. The academic community together with stakeholders from the industry and the applied practice and the research should work collaboratively on a new era of health profession education with an emphasis on the capacity of active learning to promote a holistic value proposition, integrating:

- Student-Centric Learning
- Personalized Exploration
- Orientation with Learning Objectives
- Enriched Learning Context
- Team Collaboration
- Problem Solving
- Faculty Development and Reward
- Technology Impact

In the concluding chapter of this volume, we elaborate further on the core ideas introduced in this chapter. We do hope that our edition will contribute to the body of knowledge on active learning.

References

Alsanea, N. (2012). The future of health care delivery and the experience of a tertiary care center in Saudi Arabia. *Annals of Saudi Medicine, 32*(2), 117–120.

Belliger, A., & Krieger, D. J. (2018). The digital transformation of healthcare. In *Knowledge management in digital change* (pp. 311–326). Cham: Springer.

Cresswell, K., Coleman, J., Slee, A., Williams, R., Sheikh, A., & ePrescribing Programme Team. (2013). Investigating and learning lessons from early experiences of implementing ePrescribing systems into NHS hospitals: A questionnaire study. *PLoS ONE, 8*(1), e53369.

Deloitte. (2018). *The 2018 global health care outlook: The evolution of smart health care.* Deloitte.

Gopal, G., Suter-Crazzolara, C., Toldo, L., & Eberhardt, W. (2019). Digital transformation in healthcare–architectures of present and future information technologies. *Clinical Chemistry and Laboratory Medicine (CCLM), 57*(3), 328–335.

Haggerty, E. (2017). Healthcare and digital transformation. *Network Security, 2017*(8), 7–11.

Kraus, S., Schiavone, F., Pluzhnikova, A., & Invernizzi, A. C. (2021). Digital transformation in healthcare: Analyzing the current state of research. *Journal of Business Research, 123*, 557–567.

Lytras, M. D., Papadopoulou, P., & Sarirete, A. (2020). *Smart healthcare: Emerging technologies, best practices, and sustainable policies.* Elsevier. https://doi.org/10.1016/b978-0-12-819043-2.00001-0.

Lytras, M. D., Sarirete, A., & Stasinopoulos, V. (2020). *Policy implications for smart healthcare.* The international collaboration dimension. Elsevier. https://doi.org/10.1016/b978-0-12-819043-2.00017-4.

Lytras, M. D., Serban, A. C., Ruiz, M. J. T., Ntanos, S., & Sarirete, A. (2022). *Translating knowledge into innovation capability: An exploratory study investigating the perceptions on distance learning in higher education during the COVID-19 pandemic—The case of Mexico.* Elsevier BV. https://doi.org/10.1016/j.jik.2022.100258.

Misseyanni, A., Marouli, C., Papadopoulou, P., Lytras, M., & Gastardo, M. T. (2016). Stories of active learning in STEM: Lessons for STEM education. In *Proceedings of the international conference the future of education* (pp. 232–236).

Misseyanni, A., Papadopoulou, P., Marouli, C., & Lytras, M. D. (2018). *Active learning strategies in higher education.* Emerald Publishing Limited.

Naeve, A., Yli-Luoma, P., Kravcik, M., & Lytras, M. D. (2008). A modelling approach to study learning processes with a focus on knowledge creation. *International Journal of Technology Enhanced Learning, 1*(1–2), 1–34.

Sairete, A., Balfagih, Z., Brahimi, T., Amin Mousa, M. E., Lytras, M., & Visvizi, A. (2021). *Editorial—Artificial intelligence: Towards digital transformation of life, work, and education.* Elsevier BV. https://doi.org/10.1016/j.procs.2021.11.001.

Spruit, M., & Lytras, M. (2018). *Applied data science in patient-centric healthcare: Adaptive analytic systems for empowering physicians and patients.* Elsevier BV. https://doi.org/10.1016/j.tele.2018.04.002.

CHAPTER 2

The strength of digital transition in higher education as a success factor

Rui Brito Fonseca
The School of Education and Human Development, ISEC Lisboa, Lisbon, Portugal

1 Introduction

With the emergence of the COVID-19 pandemic, higher education institutions around the world were forced to change their routines and pedagogical models. What used to be face-to-face teaching became distance learning—synchronous or asynchronous; the environment of in-person conviviality among students in the academic campus environment was replaced by ways of conviviality and communication through social networks; the direct and face-to-face dialogue between students and teachers in a classroom environment was replaced by the dialogue mediated by communication platforms; the pedagogical model of the classroom—more or less expository—gave way to a more dynamic model, using technological communication platforms and video lessons. COVID-19 was the trigger for a change that was already underway, albeit at a slower pace. This disruptive trigger came to confront us and accelerate a reality that would probably take a decade or two to develop.

It will not be disproportionate to consider that socially the 21st century began with the pandemic originated by SARS-CoV-2, in the same way that we can mark the end of the 20th century with the fall of the Berlin Wall, the Eastern Bloc implosion, and the dissolution of the Warsaw Pact. Both the pandemic that we are still experiencing and the end of the Eastern Bloc had and still have global impacts that introduced significant changes in our ways of living, communication, the economy, health, and education, that is, in almost all the dimensions of our lives. We are facing a global, simultaneous, and intense change in society. As with the fall of the Eastern Bloc and associated phenomena, we are witnessing the emergence of new models of society at a global level. Society is also changing at breakneck speed as a result of the pandemic caused by SARS-CoV-2 simultaneously across the planet. A new society is emerging, with some contours that are still somewhat diffuse and with impacts that are difficult to measure (Comissão Europeia, 2020). We live, in fact, in very interesting times.

The impact of the pandemic was also felt in higher education in quite an intense way; thus, it is important to "explore the potential of digital technologies for learning and teaching and develop digital skills for all. Education and training are fundamental for

Active Learning for Digital Transformation in Healthcare Education, Training and Research
https://doi.org/10.1016/B978-0-443-15248-1.00015-1

personal fulfillment, social cohesion, economic growth and innovation" (Comissão Europeia, 2020, p. 1). In just a few weeks, higher education institutions were forced to suspend face-to-face classroom teaching activities and replace them with distance teaching activities. To this end, they reorganized pedagogical routines and models, where adaptation to synchronous and asynchronous models was essential (Farnell, Skledar Matijevic, & Scukanec Schmidt, 2021). The pedagogical model of the traditional classroom, more expositive and with well-defined social and spatial roles, was quickly transformed into a multiple digital pedagogical environment—multiple because it is not limited to a change in the mediation of communication, multiple because in addition to digital and distance, it uses different communication platforms, social networks, and complementary digital forms of study that introduce more dynamic and interactive forms of teaching that employ applications and video lessons as complementary modes of the teaching-learning process (Aristovnik, Kerzic, Ravselj, Tomazevic, & Umek, 2020). The essential part of communication among students and between students and teachers began to take place through multiple means, transforming academic life and the ways of contact with administrative and teaching services. All or almost all dimensions of the academy moved to the digital arena, gradually—but quickly—modifying the relational dimensions of all (Farnell et al., 2021; Gabriels & Benke-Aberg, 2020).

With the pandemic originated by SARS-CoV-2 and subsequent strains, the necessary path to a more digital and environmentally sustainable society has gone from a project in gradual development to a reality in accelerated implementation. This virus and the pandemic that it originated were the trigger that launched this intense social change in progress, with impacts that are still unknown, particularly with regard to higher education. In 2 years, higher education took a two-decade technological leap.

The concept of mobility has also changed. With the digitization of higher education, it became possible to carry out digital mobilities and thousands of online scientific conferences and congresses, with real impacts on countries' economies and on their environmental sustainability, as well as on the strengthening of international interinstitutional academic relations, between geographically distant higher education institutions. Our conception of geographic space has changed as relational space has become the space of digital communication (Farnell et al., 2021; Gabriels & Benke-Aberg, 2020).

We are facing a new era where the rapid digital transition is already changing higher education, placing the student at the center of the educational process, and generating new and more fruitful educational processes, perhaps more enriching for teachers and students (Comissão Europeia, 2020).

2 Social visibilities and invisibilities

With the pandemic caused by SARS-CoV-2 and the consequent confinements that led to the rapid transition from traditional face-to-face teaching to technology-based distance learning, there were many social and digital inequalities that this process allowed to be removed from the shadows, questioning the limits of it and the necessary measures to overcome it. Above all, the central role of the education system became visible in reducing learning gaps, as well as in promoting equal opportunities in society or as a legitimate "social elevator" that enhances upward social mobility.

This fact led to the urgency of distributing more educational and digital resources to the students most in need of adapting pedagogical models and teaching-learning processes to the particularities of these same students, in order to guarantee the necessary equality of opportunities, as digital technologies can indeed significantly contribute to more inclusive and qualitatively superior education and training (Farnell et al., 2021).

With the digitization of higher education and the development of hybrid and distance education, we are facing social inequalities that are situated in the field of the digital sphere, such as the digital divide, equal access to reliable information sources, equal access to digital networks of broadband with high-quality signal, and access of these students to internet services in economically sustainable conditions. Only with the resolution of these factors of social and digital inequality will we be able to speak of an effectively inclusive higher education, generating a strong sense of belonging to their higher education institution for students, allowing full learning in conditions of digital equality. When we talk about digital equality, synthetically, we are talking about equal access to digital resources (computers, tablets, android phones, etc.), good connectivity conditions, and lifelong updating of digital skills. It is also necessary that higher education institutions use digital resources capable of enhancing and personalizing learning ("learning analytics"), in order to promote an improvement in equity and social and digital inclusion (Comissão Europeia, 2020; Prova Fácil, 2019).

With the learning carried out during the first confinements and the subsequent diagnoses of needs and ongoing reflections, it is now possible to look at distance and hybrid higher education as possible, necessary, and urgent realities for the future of higher education. As an example, it is necessary that the higher education national authorities promote and support the updating and redesign of curricula for the online format, also ensuring that higher education institutions are equipped with qualified teaching staff and technological infrastructure that allow this change (Comissão Europeia, 2020; Farnell et al., 2021).

Likewise, national authorities must continue to encourage higher education institutions to promote their internationalization and international mobilities (real or virtual), without losing quality or losing their social dimension (Farnell et al., 2021; Gabriels & Benke-Aberg, 2020). As referred by UNESCO IESALC (International Institute for Higher Education in Latin America and the Caribbean) (2020, p. 3; cited in Farnell et al., 2021, p. 60), nowadays, governments have to take measures to stimulate the economy, and higher education "must be seen as a tool in a context of economic recovery and, as such, must be an integral part of the stimulus programs that are designed".

Without this broad view of context, we will be losing opportunities to develop higher education and, consequently, our model of society that, once lost, will be hard to recover.

3 Dead end, one-way street, or hybrid path under construction?

As has happened with the means of communication throughout history, the appearance and massive development of a means of communication does not condemn its predecessor to its disappearance. On the contrary, the history of the media has shown us that they do not replace each other; they add up. When the book appeared, the flying sheets (precursors to the written press) did not disappear. When the radio appeared, people did not stop reading books and newspapers. When television appeared, the radio, books, and newspapers remained as important means of communication. And when the internet became widespread, people did not stop watching television (in fact, television itself was integrated into the internet through streaming), listening to the radio, or reading books or newspapers. Today, we all read newspapers and books, listen to the radio, watch television, and search the internet, sometimes at the same time.

Likewise, with the development of the digital economy and higher education, new professional functions are already emerging, but we will not be facing the disappearance of functions, but rather the adaptation of functions to the emerging digital realities. As such, we are in a phase of development of the digitalization of higher education in which the most important thing is to train professionals, teachers, and nonteaching staff for the new functions they will perform (Comissão Europeia, 2020; Pedro, Lemos, & Wünsch, 2011). Professional training and professional empowerment for the new challenges posed by the digitalization of education are essential for higher education institutions to be able to make the necessary leap toward distance and hybrid learning models (Comissão Europeia, 2020; Farnell et al., 2021; Leite, Lima, & Monteiro, 2009). Big data, machine learning, and cloud computing are technologies of our days that have real impacts on teaching-learning methodologies.

However, we must bear in mind that distance learning is nothing new, since, in the 19th and 20th centuries, many training courses were carried out by post. We can look at correspondence teaching, the precursor to distance education by digital means. Already in the 19th century, it had a presence in countries whose their geographic size resulted in great distances between training centers and many cities, as in the case of Brazil or the United States (Almeida, 2003). As of today, distance learning is based on an educational model whose most visible feature is the physical separation and the distance between the teacher and student (Romani, 2000). With the technological evolution and the development of digital media, especially during the aftermath of the COVID-19 pandemic, distance learning and virtual mobility gained a new relevance, taking on the role of the teaching paradigm of the future, with a wide expression already at the present moment in many higher education institutions, around the world (Gabriels & Benke-Aberg, 2020).

Higher education institutions, in the incessant search for innovation, creativity, and digital entrepreneurship—all over the world—have been reformulating their degree programs, in order to incorporate subjects as autonomous curricular units or in a transversal way in the programs such as programming, design thinking, digital culture, and digital marketing, generating more autonomous, more critical, more entrepreneurial, and more versatile students. With the digitalization of higher education that places the student at the center of learning process with a stronger study and research autonomy, there will be a repositioning of the human being in a process that never excludes him and, inversely, amplifies his relevance, as it gives him more pedagogical autonomy in managing autonomous study times, highlighting their socio-emotional skills (Almeida, 2020; Comissão Europeia, 2020; Nova & Alves, 2003).

This is a path of no return, as the technological and social leaps we have made are such that returning to the old teaching models is not only a wastage of the installed technological capacity and the pedagogical progress made in the last 2 years but also a countering of all development, technological and social, provided by the expansion and popularization of access to information and communication technologies. The entire educational system will have to adapt, starting with teachers and students, active agents in the development of new pedagogical models of teaching-learning, to be increasingly focused on students with the growing adoption of computing technologies in these processes, making learning more creative, delocalized, autonomous, and fluid (Comissão Europeia, 2020; Dencker, 1998; Gabriels & Benke-Aberg, 2020; Leite et al., 2009; Saraiva, 2008). Students, central agents of this digital transformation, will also have to understand the gains that this transformation brings to their learning processes and life. In the future, we will have more satisfied students, more involved in the teaching-learning processes, with quick access to the contents of the curricular units, quick feedback from the teachers on the items to be improved, greater ease of access to their assessments, and faster and more shared contact with their teachers. But it is also necessary to change the paradigm of administrative and bureaucratic procedures, making them more dematerialized and safer. We will increasingly have to move toward virtual academic services, with storage of all academic processes in the cloud; virtual assistance to students; systems for registering and launching applications; fully virtualized enrollments and grades (among other processes); moving staff attached to these services for telework, giving them more autonomy and a better quality of life (Farnell et al., 2021; Gabriels & Benke-Aberg, 2020). For this process to be successful, there must be an alignment of objectives and tasks and a simultaneity of dematerialization of processes between teaching and nonteaching activities, toward a digital academic management that coordinates these changes in space and time, controlling processes and flows of the different dimensions of the academic institution. The dematerialization and reduction of the bureaucracy of processes will leave teachers with more time for teaching activities and for scientific research, enabling them to focus on working with students and scientific research (Comissão Europeia, 2020).

It is clear that we are talking about a major organizational change that should take place in a short period of time, in order for there to be benefit for all those involved and for overcoming any resistance. It is clear that it is necessary to involve teachers in this process of digital and educational transformation, boosting their productivity and efficiency and freeing them up with time for other pedagogical and scientific tasks (Prova Fácil, 2019).

The pandemic, in addition to the negative impacts on the health of citizens and health systems, also allowed for the development of a broad reflection on the opportunities generated by distance learning and hybrid learning, leading to reflections on the future of the teaching profession. Naturally, we are talking about greater and continuous teacher training in digital skills and new pedagogical models, in order to allow them to produce a new type of teaching sessions (synchronous and asynchronous) so that there is no temptation to replicate the models of traditional teaching in distance learning and hybrid learning. We are also talking about a reorganization of spaces and school activities with the growing extension of national and international collaborative networks. Basically, just as distance learning and hybrid learning promote greater autonomy for students, they promote the same for teachers (Farnell et al., 2021).

Finally freed from the pedagogical and teaching-learning models of the 19th and 20th centuries, the teachers of the 21st century will have the opportunity to finally be able to focus on the students, paradoxically producing a more humanized teaching that is closer to them. With distance learning—above all—it is possible to bring more higher education opportunities to developing countries, expanding the possibilities of higher education for the global population, universalizing and democratizing it, and enhancing equity in access and frequency of higher education courses, thus contributing to the regional development of parts of the planet that would hardly have access to higher education (Benetti et al., 2006; Comissão Europeia, 2020; Farnell et al., 2021; Gabriels & Benke-Aberg, 2020; UNESCO, 1997).

We are therefore not at a dead end for higher education as the options are multiple and diverse; nor are we on a one-way street as there are several ways to achieve a successful digital transition in higher education. What we have is a hybrid path under construction, which, as we have already seen, involves all sectors of the higher education institutions, necessarily well aligned and involving all agents intervening in the processes. According to the study cycles characteristics and objectives, this digital transformation can immediately move toward a distance learning model or a hybrid model, both being able to coexist in the same higher education institutions, depending on the study cycles and the strategic management options of itself (Comissão Europeia, 2020; Farnell, 2020). We are on a locomotive in top gear whose pedagogical and institutional costs of a reverse gear would be extremely high and unsustainable.

4 Ready for the digital transition of higher education?

With regard to higher education, even after the first populations enclosures ordered to contain the pandemic, many higher education institutions are still not properly prepared for the digital transition and distance learning. Also, the formation of the higher education teaching staff, oriented toward the development of digital skills, is still somewhat dispersed, disorganized, and little diversified, contributing to the lack of training for many of these teachers at this level (Pedro et al., 2011). On the other hand, there still seems to be no minimum reference of competences that these teachers must possess so that they can exercise their professional activity with competence and rigor.

It should be noted that different levels of digital training for higher education teachers are identifiable within each country and vary from country to country. It could be said that within the framework of the European Union, it is urgent to establish minimum levels of digital training for higher education teachers that standardize training. In this sense, training programs and models for the development of the teaching career have been developed over the years, with the capability of incorporating these items into the assessment of higher education teachers themselves (Comissão Europeia, 2020; Farnell et al., 2021). This need has made it urgent to carry out systematic studies on the real use of ICT in an academic environment so that we can obtain reliable data on the real digital skills of teachers, in order to make it possible to develop diagnoses on the subject, as the quality of teaching of the present and the future largely depend on these diagnoses and the development of strategies to fill their needs (Espinosa, 2010; Georgina & Hosford, 2009; Mac Labhrainn, McDonald Legg, Schneckenberg, & Wildt, 2006; Pérez, 2009; Prendes, Castañeda, & Gutiérrez, 2011).

In fact, for the digital transition to be a reality in higher education, whether in distance learning or hybrid learning, there is a set of conditions that must be guaranteed from the outset. As already mentioned here, the training of teaching and nonteaching staff is necessary so that the virtualization of higher education becomes a reality, increasingly present and closer. With distance learning or hybrid learning, teachers have to rethink the learning objectives of each curricular unit, integrating digital technologies in learning and, in this way, redefining teaching methods so that they generate more competences, capable of giving to students full answers at the cognitive and metacognitive fields, strengthening their socio-emotional mastery of technologies and of attitudes and values. Education is increasingly focused on skills and less on qualifications, capable of responding better to changes in the labor market, technological developments, and demographic and environmental transitions (Comissão Europeia, 2020; Farnell, 2020; Farnell et al., 2021). This should be the future of higher education, with new models of teacher-student-teacher interaction that uses a greater variety of teaching methodologies, integrated into the social and economic environment of students, in order to promote growing interfaces with nonacademic community, in line with what has already been done, and it is already developed at the level of polytechnic higher education.

Of course, not everything is a benefit in this digital transition: there is no *light* without its *shadow*. If the digital transition enhances access to numerous digital learning resources, it is extremely important to ensure their quality. Likewise, digital technologies allow for a real transformation of teaching practices (virtual, real-time, experimental), may also involve some loss of experiences outside the classroom, such as peer-to-peer learning. With increasing individualized and adaptive learning through digital means of assessment, some social and digital inequalities may emerge, resulting from the existence of inequalities in access to digital technologies (especially those related to access to broadband networks). In addition, digital education (in teaching to distance learning or hybrid learning) poses new challenges related to data protection. Finally, it is very important to emphasize that distance learning in higher education allows expanding access to higher education to students who otherwise would have difficulty accessing the same, enhancing lifelong learning in a more effective way; access to higher education for students from more remote regions; and the growing virtualization of students mobilities, with direct impacts on the growing internationalization of higher education institutions (Comissão Europeia, 2020; Farnell et al., 2021; Gabriels & Benke-Aberg, 2020; Samartinho & Barradas, 2020; Santiago, 2021).

Naturally, the integration of digital technologies in learning processes requires strong public investment in facilitating access to quality computer equipment for the entire population, teachers, and higher education institutions, that is, a true technological plan that drives the digital transition (Farnell et al., 2021). At the same time, assessment agencies have to adapt the study cycle evaluation systems to this new emerging reality, defining new guidelines for monitoring and quality assurance, as well as new models for certification and validation of learning.

There is therefore still a long way to go. A path that does not end with technological means and access to them; the training of teachers and pedagogical adaptations; the adaptation of higher education institutions and their provision of more and better technological equipment; the evaluation, certification, and validation systems of the study cycles and apprenticeships; and the necessary and urgent public support for this desideratum (Comissão Europeia, 2020; Farnell, 2020). There is no future alternative to the digital transition to higher education. Of course, this does not mean that face-to-face teaching will be doomed. On the contrary, it will coexist in study cycles in which the practical and laboratory component is decisive but will be limited to a small number of higher education study cycles.

Having overcome the difficulties inherent in the complete implementation of distance learning and hybrid learning, in order to speak of a real digital transition in higher education, we will have to be able to take these bold steps; otherwise, we will be condemning a process of real social and educational change when it is still "taking its first steps". It is therefore urgent to deepen reflections and design models for implementing this necessary digital transition in higher education, in close dialogue with all educational partners, particularly with the governmental authorities and the higher education assessment agency. The success of the digital future of higher education depends on everyone.

5 Conclusion

As a result of the pandemic caused by SARS-CoV-2—which has not yet ended—we have arrived at the dawn of a new era, the era of the digital society, particularly of higher education, which marks in a very significant way the global beginning of the 21st century.

The challenges facing higher education institutions and society in general are enormous. We have giant challenges related to installed technological capacity and its territorial dispersion, with the technological capacity of the different agents involved (public and private) and universal access to networks and computer equipment, essential for this digital transition to be a carrier with social equity. But we also have very pressing challenges in adapting higher education institutions and their teaching staff to this true digital revolution, requiring more specific training for everyone.

Once here, it is counterproductive to want to stop this rapidly developing process. On the contrary, it is urgent to reflect on it, to invest in the training of higher education teachers and nonteaching staff, to involve students in the processes, and to develop important changes in the legislative body and in the evaluation models of higher education, in addition to freeing public support for all this accelerated mutation. Without significant public investment in this digital transition process, it will be more difficult to meet the challenges posed by the unstoppable development of distance education and hybrid education with quality and equity. We all need to be aware that the future of higher education has arrived and that it is certainly digital.

6 Active learning sections

1. Suggested teaching assignments
- How can digital teaching change higher education bringing more success to students?
- Can we overcome the digital divide and bring more equity to higher education through digital technologies?

2. Recommended complementary readings or references
- The Future of Universities: The Digital Campus
- Digital Universities – 2030: Quel nouveau monde pour la formation et l'emploi?
- How Digital Transformation is Changing the Way Universities Operate
- EODLW 2021: Practices in Digital Education for Universities
- The Era of Online Learning | Niema Moshiri | TEDxUCSD

3. Case study

Currently, among other academic and management tasks that I have, I have taught curricular units in post-graduate courses. These are 100% online postgraduate courses whose general objectives are to train professionals and students for conceptual and practical field intervention.

The great novelty of this post-graduate courses is that it comprises a limited number (nearly one-fourth of total time) of contact hours (online classes and tutorials) and a large number of hours of autonomous work (nearly three-fourths of total time) for students, which allows them to reflect on the themes in study in a more mature way. This is autonomous work that does not consist of solitary work. On the contrary, we are dealing with autonomous work built in a network, at a distance, with other colleagues and with the teacher. As such, it is an extremely shared work among all students, much more enriching and generating more synergies than the work carried out in the classroom, as by enabling and enhancing the discussion of ideas between students from different parts of the country or the world, with more differentiated academic, personal, and professional life experiences, it greatly enriches the teaching-learning process.

In addition, the attention that can be captured from students in a digital environment is far superior to that achieved in the classroom, as the digital classroom allows for greater interaction between students and the teacher and the construction of more fluid learning processes.

The experience lived in teaching this curricular units allowed me to recognize in practice the clear advantages of distance learning by digital means. From the students' point of view, as expressed by themselves, the experience was similarly rich. According to them, the fluid and open way in which the classes took place, the flexibility of learning, and the autonomy of knowledge acquisition processes carried out in a network were assets that positively marked their learning processes.

Based on this experience, can we state that e-learning is the future of higher education?

What about more practical study cycles? Is the solution b-learning, or should they stay presencial?

Are we building different levels of higher education recognition, given the multiplicity of existing study cycles?

4. Titles for research essays

- Can online higher education teaching bring more opportunities for students from less developed countries?
- Are we entering a new golden era of autonomous learning through e-learning and b-learning?
- E-learning or b-learning? A hard choice to be made.

5. Recommended Project URLs for further research

- Digital transformations in higher education
- https://educationaltechnologyjournal.springeropen.com/articles/10.1186/s41239-021-00287-6
- https://elearningindustry.com/digital-transformation-in-higher-education-8-top-trends
- Digital Education Action Plan (2021-2027) - Education and Training - European Commission

References

Almeida, M. E. (2003). *Educação e distância no Brasil: diretrizes políticas, fundamentos e concepções*. Campinas: Papirus.

Almeida, C. (2020). Os efeitos da cultura digital no ensino superior. In *Revista do Ensino Superior*. 24 de abril de 2020. disponível em https://revistaensinosuperior.com.br/cultura-digital-ensino-superior/. (Accessed 31 October 2021).

Aristovnik, A., Kerzic, D., Ravselj, D., Tomazevic, N., & Umek, L. (2020). Impacts of the COVID-19 pandemic on the life of higher education students: A global perspective. *Sustainability*, *12*(20), 8438. Agosto de 2020 disponível em https://www.researchgate.net/publication/343555357_Impacts_of_the_COVID-19_Pandemic_on_Life_of_Higher_Education_Students_A_Global_Perspective. (Accessed 1 November 2021).

Benetti, K., Melo, P., Spanhol, F., Pacheco, A., Dalmau, M., & Tosta, H. (2006). *A formação docente no Brasil, na América Latina e no Caribe*. Florianópolis: Paper Print.

Comissão Europeia. (2020). Plano de Ação para a Educação Digital 2021-2027 – Reconfigurar a educação e a formação para a era digital. In *Comunicação da Comissão ao Parlamento Europeu, ao Conselho, ao Comité Económico e Social Europeu e ao Comité das Regiões*. Bruxelas: Comissão Europeia.

Dencker, A. (1998). *Métodos e técnicas de pesquisa em turismo*. São Paulo: Futura.

Espinosa, M. (2010). Competencias TIC para la docência en la Universidad Publica Española: Indicadores e propuestas para la definición de buenas prácticas - Programa de Estudio e Analisis. In *Informe del proyecto EA2009-0133 de la Secretaria de Estado de Universidades e Investigación*. Madrid: Secretaria de Estado de Universidades e Investigación.

Farnell, T. (2020). *Community engagement in higher education: Trends, practices and policies*. NESET report Luxembourg: Publication Office of the European Union. https://doi.org/10.2766/64207.

Farnell, T., Skledar Matijevic, A., & Scukanec Schmidt, N. (2021). *The impact of COVID-19 on higher education: A review of emerging evidence – Analytical report*. NESET report Luxembourg: Publication Office of the European Union. https://doi.org/10.2766/069216.

Gabriels, W., & Benke-Aberg, R. (2020). *Student exchanges in times of crisis – Research report on the impact of COVID-19 on student exchanges in Europe*. Erasmus student network: AISBL.

Georgina, D. A., & Hosford, C. C. (2009). Higher education faculty perceptions on technology integration and training. *Teaching and Teacher Education: An International Journal of Research and Studies*, *25*(5), 690–696.

Leite, C., Lima, L., & Monteiro, A. (2009). O trabalho pedagógico no ensino superior. Um olhar a partir do prémio de excelência e-learning da Universidade do Porto. In *vol. 28. Educação, Sociedade & Culturas* (pp. 71–91). Porto: Universidade do Porto.

Mac Labhrainn, I., McDonald Legg, C., Schneckenberg, D., & Wildt, J. (Eds.). (2006). *The challenge of ecompetence in academic staff development*. NUI Galway: CELT.

Nova, C., & Alves, L. (2003). Educação a Distância: limites e possibilidades. In *Educação a Distância: uma nova concepção de aprendizado e interatividade*. São Paulo: Futura.

Pedro, N., Lemos, S., & Wünsch, L. (2011). E-learning no Ensino Superior: benefícios e limites na perspetiva dos estudantes. In *Actas do challenges 2011 – VII international conference on ICT in education*. Braga: Universidade do Minho.

Pérez, K. (2009). La competencia digital del professorado universitário para la sociedad del conocimiento: un modelo para la integracion de la competencia digital en el desarollo professional docente. In *V Congreso de Formación para el Trabajo. Granada* (pp. 24–27).

Prendes, P., Castañeda, L., & Gutiérrez, I. (2011). University teachers ICT competence: Evaluation indicators based on a pedagogical model. In *Educação, Formação e Tecnologias, nº extra, Abril 2011* (pp. 20–27). Monte da Caparica: Universidade Nova de Lisboa.

Prova Fácil. (2019). *Por que a transformação digital no Ensino Superior é um caminho sem volta?*. 14 de fevereiro de 2019. disponível em https://www.provafacilnaweb.com.br/blog/transformacao-digital-no-ensino-superior/. (Accessed 31 October 2021).

Romani, L. (2000). *InterMap: Ferramenta para visualização da interação em ambientes de Educação a Distância na Web*. Campinas: ICUEC.

Samartinho, J., & Barradas, C. (2020). "Editorial: A Transformação Digital e Tecnologias da Informação em tempo de Pandemia". Conferência Virtual A Transformação Digital e Tecnologias em Tempo de Pandemia. *Revista da UIIPSantarém. Edição Temática: Ciências Exatas e Engenharias, 8*(4), 1–6. disponível em https://revistas.rcaap.pt/uiips/. (Accessed 29 October 2021).

Santiago, P. (2021). Desafios actuais para os Sistemas Educativos – Uma perspectiva de política pública da OCDE. In *Que modelo de educação queremos para o futuro?* OCDE. Seminário virtual. 17 de março de 2021.

Saraiva, T. (2008). Educação a distância no Brasil: lições da história. In *Em Aberto. ano 16. Nº 70*. Brasília: INEP.

UNESCO. (1997). *Aprendizagem aberta e a distância: perspectivas e considerações políticas educacionais*. Florianópolis: Imprensa Universitária.

CHAPTER 3

Support of educational videos to improve the knowledge of health professionals

Sandra Laia Esteves[a,b,c,d] and Ana Veiga[b,e]
[a]Regional Health Administration of Lisbon and Tagus Valley
[b]Fiscal Council of the Portuguese Society of Health Literacy
[c]Health Literacy from ISPA
[d]Management in Health
[e]Dr. Gama Institute of Ophthalmology; Escola Superior de Enfermagem de Lisboa (ESEL)

1 Strategies facilitating communication in health

Health communication encompasses the interactions between health professionals and users as a dynamic process that, when effective and functional, promotes decision-making and concomitantly produces positive health outcomes (U.S. Department of Health and Human Services, 2000).

The Directorate General of Health (DGS) (2017) emphasizes that communication in health requires knowledge, competence, and empathy. Thus, the healthcare professional must know when to speak, what to say, and how to say it. Although communication is used daily during healthcare delivery, it requires skills that must be learned and practiced in order to contribute to establishing effective communication in dynamic environments common to healthcare professionals. At the same time, Colorado Health Outcomes Program (CHOP) (2015) points out that for effective and functional communication, it is important to prepare health professionals in a pedagogical educational approach to provide user-centered care as members of a multiprofessional team.

The application of communication-facilitating strategies promotes safety in communicating clearly and effectively in the relationship that the health professional establishes with the user, which will benefit users to understand information related to their health and, at the same time, feel more involved in their health process (Fig. 1).

Crucial to this is the simultaneous use of improved communication techniques (Fig. 2) such as the teach-back method, in which it is important for healthcare professionals to ensure that clients have understood the information provided to them during care. Follow-up is the act of contacting a client or caregiver after the last visit to assess the client's condition, identify misunderstandings, clarify questions, and, if necessary, adjust treatment. The brown bag review method allows you to identify medical errors and misunderstandings, review, and confirm with patients "how" and "when" to take their

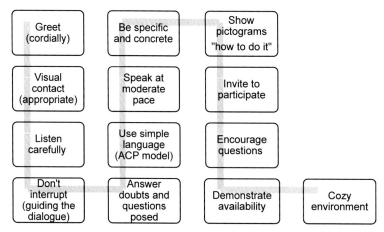

Fig. 1 Strategies facilitating communication in health. *(Source: Esteves, S. L., & Lopes, A. S. (2021). Handbook of research on assertiveness, clarity, and positivity in health literacy. Available from: https://www.igi-global.com/book/assertiveness-clarity-positivity-health-literacy/273602 (Accessed November 2021), based on Almeida, C. (2020).* The contribution of communication skills of doctors and nurses to health literacy: The ACP model—Assertiveness (A), Clarity (C) and Positivity (P) in the therapeutic relationship. *Available from: https://www.repository.utl.pt/handle/10400.5/20901. [Accessed March 2021] and Colorado Health Outcomes Program. (2015). AHRQ health literacy universal precautions toolkit. Aurora: Agency for Healthcare Research and Quality.)*

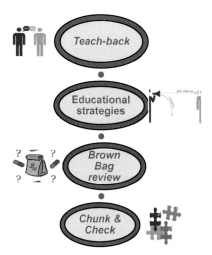

Fig. 2 Techniques for improving communication. *(Source: Esteves, S. L., & Lopes, A. S. (2021). Handbook of research on assertiveness, clarity, and positivity in health literacy. Available from: https://www.igi-global.com/book/assertiveness-clarity-positivity-health-literacy/273602 (Accessed November 2021), based on Almeida, C. (2020).* The contribution of communication skills of doctors and nurses to health literacy: The ACP model—Assertiveness (A), Clarity (C) and Positivity (P) in the therapeutic relationship. *Available from: https://www.repository.utl.pt/handle/10400.5/20901. [Accessed March 2021]; Colorado Health Outcomes Program. (2015). AHRQ health literacy universal precautions toolkit. Aurora: Agency for Healthcare Research and Quality; National Health Service (2020). Chunk & Check. The Health Literacy Place. Available from: http://www.healthliteracyplace.org.uk/tools-and-techniques/techniques/chunk-and-check/. [Accessed March 2021].)*

medication in a safe and appropriate manner. The provision of written information (for example, leaflets) is another educational strategy that should be discussed with the user at the time of delivery; it is crucial to underline or highlight the most important points and personalize the leaflet. Encouraging questions is fundamental, as it is well known that most users feel inhibited when asking questions. Creating a welcoming environment that counteracts this can lead to users feeling like active partners in their Health process (CHOP, 2015). The chunk and check is another crucial method in the professional/user relationship because it allows the division of the information into "blocks" and is progressively transmitted from the simplest to the most complex. It requires checking the user's understanding of his health process and also correcting misunderstandings about facts or misconceptions. The health professional issues the information and summarizes and repeats the same information, in order to facilitate the user's memorization and understanding. In this process, the user is asked, by means of the teach-back method, to demonstrate that he or she can decode and understand the information provided (National Health Service, 2020).

We emphasize that the use of functional and effective communication by the health professional with the user will contribute to better adherence and behavioral changes for the benefit of their Health and Safety.

2 Training of health professionals

The empowerment of health professionals in the development of educational strategies and interaction and communication with users promotes the improvement of health literacy levels and the acquisition of greater autonomy and critical reflection in decision making for their health. The Assertiveness, Clarity, and Positivity (ACP) model can be considered an excellent resource that promotes the empowerment of professionals in the acquisition of communication skills that favor access to information, the understanding of the content, and the use of information in the users' (correct) decision-making for their health, according to principles of the Health Literacy (Almeida, 2020; Sorensen et al., 2012).

The application of the ACP Model promotes that health professionals are better able to communicate, coordinate, and optimize the therapeutic relationship with their users (Fig. 3) (Almeida, 2020).

Digital health not only embraces technological transformation but also fosters the therapeutic relationship between health professionals and users and the circumstances of the health process.

Empowering health professionals favors the use of digital technologies in their practice with greater ease, by technologies that support and develop their work and in turn empower their users in acquiring and understanding the knowledge of the best, most reliable, and trustworthy digital health sources and technologies. Empowered health professionals can also be described as "engaged" when they understand the feelings and views of their users, providing relevant feedback and involving them throughout

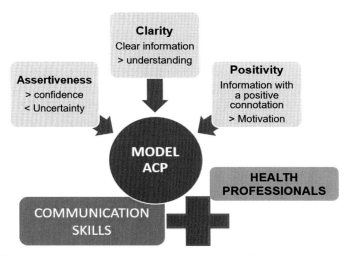

Fig. 3 The ACP model as a contribution to the empowerment of health professionals. *(Source: Own preparation, based on Almeida, C. (2020). The contribution of communication skills of doctors and nurses to health literacy: The ACP model—Assertiveness (A), Clarity (C) and Positivity (P) in the therapeutic relationship. Available from: https://www.repository.utl.pt/handle/10400.5/20901. [Accessed March 2021].)*

the healthcare process. The literature extols that the diffusion of digital health literacy recreates positive change in the health professional from a "rule follower" to a change from a creative professional, from "self confident" to being more curious and innovative, which in turn increases teamwork, namely motivation and cohesion (Fig. 4) (Gyõrffy & Mesko, 2019).

It is crucial to invest in the training of health professionals, in order to use and respond favorably to the immediate health needs of the user. It is considered a priority area, since results show that half of the European population is unable to access, understand, evaluate, and use health information properly (Sorensen et al., 2012).

It is up to health organizations, with the support of policy makers, to develop formative strategies in order to provide supportive environments and resources for health literacy interventions, which will provide the achievement of positive health outcomes.

Digital health literacy initiatives by training and motivating health professionals to understand the benefits of the correct use of digital tools create trusted solutions that facilitate the interaction between users and health services while promoting health professional-user therapeutic relationships. Thus, users are not only more informed about their health but are also driven to better skills to manage their own health (Direção-Geral da Saúde, 2019a, 2019b).

We can add that, in this way, we can see an autonomous participation of users in health-related issues, the adoption of preventive attitudes, better health choices, and, consequently, a better use of health services (Direção-Geral da Saúde, 2019a, 2019b).

In this context, user education aims to improve the knowledge and skills of individuals and their families, so as to influence the attitudes and behaviors that are essential for

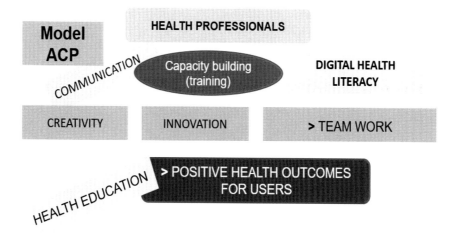

Fig. 4 Empowering health professionals in digital health literacy. *(Source: own elaboration, based on Almeida, C. (2020).* The contribution of communication skills of doctors and nurses to health literacy: The ACP model—Assertiveness (A), Clarity (C) and Positivity (P) in the therapeutic relationship. *Available from: https://www.repository.utl.pt/handle/10400.5/20901. [Accessed March 2021]; Győrffy, Z., & Mesko, B. (2019).* The rise of the empowered physician in the digital health era: Viewpoint. *Available from: https://www.jmir.org/2019/3/e12490/. [Acessed October 2021].)*

health maintenance and promotion (Bastos & Ferrari, 2011). In fact, health education is a set of knowledge and practices aimed at preventing various pathologies and a strategy to achieve the users' active participation in the management of their health-illness process (Meirelles, Sena, Cruz, Luedy, & Junior, 2013). It is understood as an integral part of treatment, as the focus is on the person rather than the disease, and allows the user to make informed choices in the management of their health condition (WHO, 2006). The benefits of user education are numerous, as there is an improvement in the following aspects: (i) communication between the user and health professionals; (ii) knowledge about their disease; (iii) participation in treatment decisions; (iv) self-care; (v) self-criticism; (vi) emotional well-being; and (vii) quality of life and use of existing resources (Reed, 2010).

3 Stages of education

The education process comprises five stages. The first consists of an assessment of the user's prior knowledge, learning abilities and styles, attitudes, cognitive level, and motivation. Subsequent to this assessment, resources, barriers, and learning needs are identified. The third stage is dedicated to planning education by involving the user in goal setting and intervention choices. The fourth step concerns the effective implementation of health education and finally evaluation (Rankin, 2001).

Education aimed at self-care in chronic diseases is called therapeutic education (Sociedade Portuguesa de Diabetologia [SPD], 2018). It is an ongoing process that consists of helping the person to develop the necessary skills to adapt their daily life to their new life condition Sociedade Portuguesa de Diabetologia (2018). It has its focus on the person and involves learning regarding self-care, psychosocial support, prescribed treatment, and health- and disease-related behaviors (WHO, 1998). Education of this nature is carried out by health professionals who accompany the person and family members throughout their life cycle with the disease.

The skills to be developed by the professionals who provide care using this intervention in their clinical practice call for some specific communication skills used in the development of a better health literacy, including empathy, active and reflective listening, congruence between verbal and nonverbal communication, valuing the active participation of people throughout the process, assertiveness, clarity of language, and positive encouragement (Almeida, 2019).

The construction of a therapeutic education program is based on four phases, in which the health professional has different skills Sociedade Portuguesa de Diabetologia (2018), as shown in Fig. 5.

Interventions using therapeutic education should use an approach that meets the needs and limitations of the user and encourages changes in behavior and self-care. Thus, there should be a participatory dynamic supported by four principles (Fig. 6).

Toward the therapeutic education, a strategic approach for the provision of trusted information is required. In this context, strategic goals, decision-making processes,

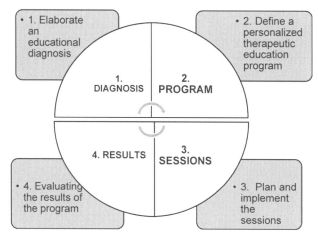

Fig. 5 Phases of the therapeutic education program. *(Source: Own preparation, based on Sociedade Portuguesa de Diabetologia. (2018).* Therapeutic education in diabetes – Competencies of health professionals and people with diabetes-2018. *SPD. http://spd.pt/images/booklet_educação_terapeutica.pdf.)*

POWER	MOTIVATION
Delegating authority and responsibility, which means giving importance and trust to the person, giving him/her freedom and autonomy of action	Continuous encouragement, which means recognizing good performance, highlighting results, and praising the achievement of goals

DEVELOPMENT	LEADERSHIP
Provide resources in terms of training and personal development. Continuous training, providing information and knowledge	Providing guidance, which means setting objectives and goals, opening new horizons, evaluating performance, and providing feedback

Fig. 6 Principles of therapeutic education. *(Source: Own preparation, based on Sociedade Portuguesa de Diabetologia. (2018). Therapeutic education in diabetes – Competencies of health professionals and people with diabetes-2018. SPD. http://spd.pt/images/booklet_educação_terapeutica.pdf.)*

and overall performance can be utilized jointly through overarching principles and strategy formulation. This requires (Fig. 7):

Therapeutic education opens space for the reformulation of the traditional professional–patient relationship, giving special focus to the need to center health care on the person and not on the disease (Bastos & Ferrari, 2011).

In this scenario, it is essential that health professionals are trained to act in the most appropriate way possible in terms of effectiveness, efficiency, and productivity throughout their professional career, which will translate into health gains for both the patient and the health systems.

SHARING INFORMATION
Information is essential for decision making. It must be clear, transparent and adapted to the condition and needs of the elderly person with diabetes.

RESPECT AUTONOMY
Give the person not only the information, but the support and freedom to decide and act. Errors must be corrected, not punished, because a punitive culture impedes autonomy. The person should be guided by their goals and objectives.

Fig. 7 Strategies for informing the patient. *(Source: Own preparation, adapted from Mulcahy, K., Maryniuk, M., Peeples, M., Peyrot, M., Tomky, D., Weaver, T., & Yarborough, P. (2003). Diabetes self-management education core outcomes measures. The Diabetes Educator, 29(5), 768–803. https://doi. org/10.1177/014572170302900509.)*

4 Digital health literacy: Revolutionize the way users achieve higher levels of health knowledge

In health, digital literacy refers to the ability to use, explore, understand, and critically evaluate health information provided by any digital tool (Direção-Geral da Saúde, 2019a).

For the eHealth Literacy Framework (eHLF), digital health literacy integrates the skills and resources needed for users to use and benefit from digital health resources (Norgaard et al., 2015).

According to this model, digital health literacy integrates seven domains (Fig. 8).

The World Health Organization (WHO) (2020) states that digital health solutions can revolutionize the way users achieve higher levels of health and access services in order to promote and protect their health and well-being. Advances in digital health literacy provide new challenges to healthcare organizations, health professionals, and users alike.

The symbiosis of digital health literacy and health outcomes integrates the possibility of empowering the user to be a more active participant in his or her health process. Digital educational strategies promote the provision of multimedia education, such as audiovisuals, in multiple languages, allowing the user more empowerment in their relationship with their healthcare professional/healthcare organization (Scott, 2019).

Studies report numerous benefits but also barriers to the adoption of digital health in clinical practice (Fig. 9).

However, scientific evidence shows that the perception and use of digital literacy in health is still limited, despite technological advances over time, and requires investment in order to achieve positive health outcomes for the user and the health systems themselves (Buntin, Burke, Hoaglin, & Blumenthal, 2011). It is well known that any citizen

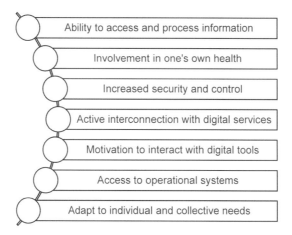

Fig. 8 Domains that facilitate digital health literacy. *(Source: Own elaboration, based on Norgaard, O., Furstrand, D., Klokker, L., Karnoe, A., Batterham, R., Kayser, L., et al. (2015). The e-health literacy framework: A conceptual framework for characterizing e-health users and their interaction with e-health systems.* Knowledge Management & E-Learning 7(4), *522–40.)*

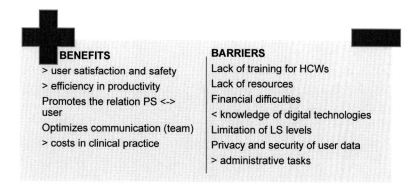

Fig. 9 Benefits and barriers to implementing digital literacy in health. *(Source: Own preparation, based on Györffy, Z., Mesko, B. (2019). The rise of the empowered physician in the digital health era: Viewpoint. Available from: https://www.jmir.org/2019/3/e12490/. [Acessed October 2021].)*

uses digital technology more frequently to obtain information or clarify questions that arise in everyday life (mostly technical jargon terms) related to health and disease. However, the resources and information collected on health are only useful if citizens are able to understand and use them properly, and the tools to evaluate this success are still very scarce (Norman & Skinner, 2006).

Thus digital health applications can exacerbate health inequalities if these applications do not meet the needs of disadvantaged populations, such as effective access to digital technologies. It is clear that users living with socioeconomic disadvantages (low income, low education, unemployment, lower levels of general and digital literacy) use digital health applications less frequently. For the authors, accessibility to the digital world is considered to be limited or compromised (Fig. 10) (Thomas et al., 2019).

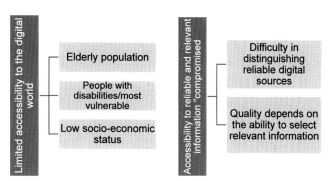

Fig. 10 Accessibility to digital health literacy. *(Source: Own elaboration, based on Direção-Geral da Saúde. (2019a). Manual of good practices health literacy. Empowering health professionals. https://doi.org/10.13140/RG.2.2.17763.30243. Available from: https://www.dgs.pt. [Accessed November 2021]; Direção-Geral da Saúde. (2019b). Plano de Ação para a Literacia em Saúde. Lisboa: Direção-Geral da saúde. Available from: https://www.dgs.pt. [Accessed December 2021].)*

The use of digital health literacy with the support and collaboration of health professionals is considered a powerful tool to bring health-related information to users with low levels of health literacy, promoting their access to and greater confidence in these digital applications (Eysenbach, 2001; Mackert, Kahlor, Tyler, & Gustafson, 2009). The US Department of Health and Human Services (2015) predicts that digital health literacy will benefit the quality of health care, namely, the user being more participatory and a manager of his or her own health.

This type of health approach has a high potential to improve access to healthcare by offering better integration and personalization of care (Latulippe, Hamel, & Giroux, 2017).

The evolution of digital literacy in health emerges as a strategy to improve the health of all, as an opportunity in the areas of disease prevention, protection, and health promotion, adapted to the needs of users, in order to prevent the increase of health inequalities. To this end, the implementation of certain strategies is recommended (Fig. 11).

Technological advances provide that digital health literacy promotes increased participation rates due to its convenience for remote users, allowing effective health guidance without the added need, in certain situations, for face-to-face attendances (Free et al., 2013; Ministry of Health, 2013).

Information and Communication Technologies (ICT) have been identified as particularly relevant tools when they are associated with the promotion of health literacy and citizens' empowerment and act as sets of technological resources that integrate and promote communication channels that favor learning (Tang, Ash, Bates, Overhage, & Sands, 2006). In Portugal, these have been identified as pathways for healthcare improvement and have been increasingly important in promoting access to healthcare and health self-management (Nijland, van Gemert-Pijnen, Boer, Steehouder, & Seydel, 2018).

As technologies evolved, Mayer sought to find methods and models that would aid their use in practice more effectively, such as the Cognitive Theory of Multimedia Learning (TCAM). For Mayer, people deepen their knowledge more from pictures and words than

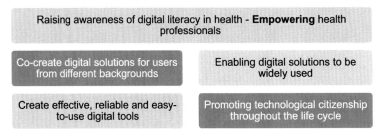

Fig. 11 Strategies facilitating digital health literacy. *(Source: own elaboration, based on Direção-Geral da Saúde. (2019a). Manual of good practices health literacy. Empowering health professionals. https://doi.org/10.13140/RG.2.2.17763.30243. Available from: https://www.dgs.pt. [Accessed November 2021]; Direção-Geral da Saúde. (2019b). Plano de Ação para a Literacia em Saúde. Lisboa: Direção-Geral da saúde. Available from: https://www.dgs.pt. [Accessed December 2021].)*

PRINCIPLES	ADVANTAGES
Multimedia	Usage: words and images
Temporal and Spatial	Words and images displayed simultaneously
Modality	Animation with narration
Redundancy	The use of graphics and narration
Coherence	Use of material and relevant information
Signaling	Flagged text (bold, italic, colors, etc)
Customization	Colloquial style information (key-words)
Segmentation	Segment information

Picture 1 Principles of multimedia learning. *(Source: Esteves, S. L., & Lopes, A. S. (2021). Handbook of research on assertiveness, clarity, and positivity in health literacy. Available from: https://www.igi-global.com/book/assertiveness-clarity-positivity-health-literacy/273602 (Accessed November 2021), based on Mayer, R. E. (2014). The Cambridge handbook of multimedia learning. New York: Cambridge University Press. Available from: https://assets.cambridge.org/97811070/35201/frontmatter/9781107035201_frontmatter .pdf. [Accessed February 2021].)*

from words alone. Multimedia representations enhance and produce more meaningful learning when the senses are combined. The author emphasizes that learning requires active processing based on dual channels (visual-pictorial and auditory-verbal), each of which has a limit to its capacity to process information. He adds that multimedia is the association of textual presentation (spoken or printed texts) with images (photos, illustrations, animations or videos) and multimedia learning is the mental construction that the learner carries out by linking words and images. So when an image, video, picture, or text is presented to the eye, this information is processed by the visual channel. If the information comes in the form of narrations or nonverbal sounds, the information is processed in the auditory channel. The human being has a limited capacity for processing information that can be processed simultaneously in the auditory and visual channels.

To make multimedia learning more efficient, Mayer suggests the integration of eight principles (Picture 1).

According to Mayer's theory (2014), the use of these cognitive principles favors digital learning, which is seen as one of the most efficient educational strategies in health education.

5 Educational strategy: Follow an audiovisual pedagogical technique

Mayer demonstrated through the Cognitive Theory of Multimedia Learning (TCAM) that the use of audiovisual pedagogical techniques (APTs) contributes to the determinant teaching process in health. This theory emphasizes the possibility of TAP to simultaneously present different sensory channels (verbal–auditory and visual–pictorial) as a robust educational strategy (Mayer, 2014).

TAP integrates a form of communication that synchronizes sound and image, using language to generate meaning (Infopédia, 2021). This technique can act as an integrative bridge between art (production) and specific knowledge, as it stimulates reflection, creativity, and critical thinking (Mendonça, Ferreira, & Rodriguez, 2014).

Scientific evidence validates that the use of TAP incites learning through its emotional component (Arroio & Giordan, 2006). It is found that with the application of TAP, the percentages of use of the five senses in terms of perceptual relevance are 83% and 11% for vision and hearing, respectively (Mayer, 2014).

TAP, due to its many advantages, is an excellent pedagogical strategy to be applied to both users and health professionals as recipients (Fig. 12).

The elaboration of the TAP should be based on certain stages: Preproduction, production, and postproduction (Fig. 13).

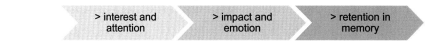

Fig. 12 Advantages of using the Audiovisual Pedagogical Technique. *(Source: Esteves, S. L., & Lopes, A. S. (2021).* Handbook of research on assertiveness, clarity, and positivity in health literacy. *Available from: https://www.igi-global.com/book/assertiveness-clarity-positivity-health-literacy/273602 (Accessed November 2021), based on Mayer, R. E. (2014).* The Cambridge handbook of multimedia learning. *New York: Cambridge University Press. Available from: https://assets.cambridge.org/97811070/35201/frontmatter/ 9781107035201_frontmatter.pdf. [Accessed February 2021].)*

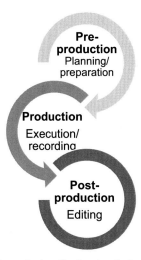

Fig. 13 Stages of elaboration of pedagogical audiovisual technique. *(Source: Esteves, S. L., & Lopes, A. S. (2021).* Handbook of research on assertiveness, clarity, and positivity in health literacy. *Available from: https://www.igi-global.com/book/assertiveness-clarity-positivity-health-literacy/273602 (Accessed November 2021), based on Leite, J. (2019).* Quais são as etapas de um projeto audiovisual? *Available from: https:// videopedia.com.br/geral/pre-producao-de-videos/ [Accessed April 2021].)*

In order to promote good practices that foster safety in communicating and promote health literacy, Lopes, Filipe, and Esteves (2020) created the TAP, entitled "E-health/Communication Safety - A Pedagogical Guidance for Health Professionals" that aims to empower and raise awareness among future and future health professionals about the importance of safety in communicating in favor of patient safety.

These digital instruments are pedagogical in nature and are intended for all professionals and students in the health area. They portray three practical problems/situations, which are transversal in health contexts. Interactions between health professionals and patients were simulated, filming short scenes in a consultation environment, namely "Doctor-Patient: Adherence to Treatment"; "Nurse-Patient: Management of Therapeutic Regime"; and "Clinical Secretary-Patient: Attendance at the Consultation" (Fig. 14).

In each of them, two scenarios are shown, the use of dysfunctional and functional communication between health professionals (Physician, Nurse, and Clinical Secretary) and the patient.

In summary, there are pedagogical guidelines advised to be applied in health communication, in order to promote safety in communicating (Lopes et al., 2020).

The development of any TAP should focus on the purpose of the instrument, in order to obtain a result that meets the author's expectations, as well as the identified needs.

Fig. 14 Pedagogical video scenes.

6 Methodology

Based on an integrative literature review in several scientific databases, we justify the relevance of the use of digital tools in health and present the results of the application of this TAP to health professionals, associated with a knowledge monitoring tool in the area of health communication.

The authors also present an audiovisual instrument for training health professionals through contrast practices made with scenarios that present, in role-play, the correct and incorrect practice. We gathered a set of interventions and produced the videos according to the principles of health literacy, namely, rapport, active listening, chunk and check, ACP model (assertiveness, clarity, and positivity), Show me. Role-plays were analyzed and evaluated by professionals who validated them as useful instruments of health education and literacy.

This TAP was presented to several multiprofessional health teams, both in an academic context (to specialization course and master's students) and in the context of primary health care (in a community care unit and in a family health unit, in an in-service training context) in the geographic area of Lisbon and Tagus Valley.

During this presentation, the authors considered it pertinent to apply a monitoring instrument (pre- and postsession) in order to assess the trainees' knowledge about the importance of health literacy, the use of functional communication strategies, and the main techniques to improve communication (Lopes et al., 2020).

This knowledge monitoring instrument is composed of two parts: the first part addresses the sociodemographic characterization, namely, age, gender, profession, and academic qualifications, and the second part integrates four multiple choice questions and one matching question, addressing the importance of "Acquisition of Skills in Health Literacy"; "Safety Strategies in Health Communication"; "Communication Enhancement Techniques"; and, finally, the frequency of "Use of Communication Enhancement Techniques" in the clinical practice of these health professionals.

To facilitate and clarify the perception of the results obtained when applying this instrument, we chose to code each questionnaire before and after the session (according to the number of trainees per presentation).

The TAP described in this methodology (translated version) is available at the following link: https://youtu.be/DEsiGK_eiAo.

7 Results and discussion

With the application of this knowledge monitoring instrument, the following results were analyzed before and after the session. Regarding the sociodemographic characterization, based on age, gender, and professional category, we found that all the technical assistants are female, aged between 46 and 50 years. The age range of the professional category nurse is between 25 and 45 years old, with 29 females and 21 males. The age range of the medical professional category is between 31 and 50 years, five of whom are female and seven male (Graphic 1) (Esteves & Lopes, 2021).

The target population of this sample includes 50 nurses, 12 physicians, and 3 technical assistants. As for the academic qualifications of these health professionals, we found that the technical assistants have completed the 12th grade of schooling, approximately 40, 1 have a degree in nursing and medicine, respectively; 9 nurses and 8 doctors have a degree, and 1, 3 have master's degrees in Nursing and Medicine, respectively (Graphic 2) (Esteves & Lopes, 2021).

The first question of the monitoring instrument is, "Does the acquisition of competencies in health literacy influence decision-making in health; change attitudes and behaviors in health; and promote the use of appropriate and functional communication strategies," which 31, 14, and 5 answered 33%, 66%, and 100% correctly in the presession, respectively. It was verified with the application of the knowledge assessment questionnaire in the presession the medical professional category answered about 66% of correct answers, in the postsession all answered correctly, so there was knowledge acquisition. The technical assistants, in the presession, answered about 33% correctly, and in the postsession, they all answered correctly. With the results obtained, we found that there were doubts about "the influence of the acquisition of skills in Health Literacy" in the presession, whereas in the postsession, there was an improvement in the results and a gain in knowledge (Graphic 3) (Esteves & Lopes, 2021). Studies have shown that

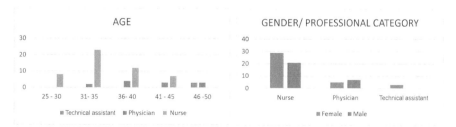

Graphic 1 Age and gender/professional category. *(Source: Esteves, S. L., & Lopes, A. S. (2021).* Handbook of research on assertiveness, clarity, and positivity in health literacy. *Available from: https://www.igi-global.com/book/assertiveness-clarity-positivity-health-literacy/273602 (Accessed November 2021).)*

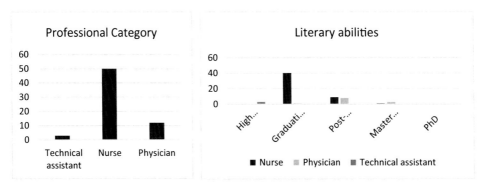

Graphic 2 Professional category and literary abilities. *(Source: Esteves, S. L., & Lopes, A. S. (2021). Handbook of research on assertiveness, clarity, and positivity in health literacy. Available from: https://www.igi-global.com/book/assertiveness-clarity-positivity-health-literacy/273602   (Accessed November 2021).)*

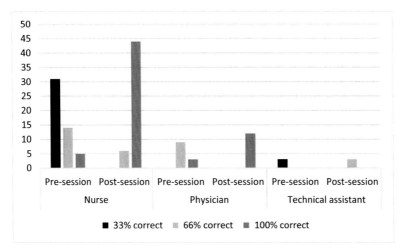

Graphic 3 Acquisition of health literacy skills. *(Source: Esteves, S. L., & Lopes, A. S. (2021). Handbook of research on assertiveness, clarity, and positivity in health literacy. Available from: https://www.igi-global. com/book/assertiveness-clarity-positivity-health-literacy/273602 (Accessed November 2021).)*

users with low health literacy have greater difficulty in understanding their health problems and making the right health decisions; make more mistakes when taking medications; use health services inadequately, including emergency care; and have a more critical health status, which leads to higher health costs (CHOP, 2015).

The second question addresses some of the "Safety Strategies for Communicating," such as the type of language to be used (use simple language, avoid the use of technical jargon, speak at a moderate pace, and prioritize what should be discussed) and the relationship between the health professional and the patient (cordial greeting, active listening and empathy). We can see in Graphic 4 that in the postsession in all professional

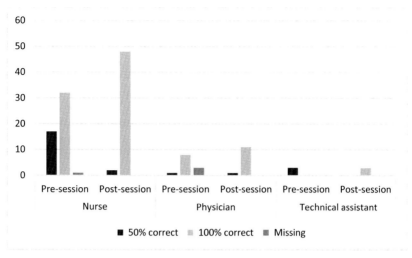

Graphic 4 Safety strategies for communicating. *(Source: Esteves, S. L., & Lopes, A. S. (2021).* Handbook of research on assertiveness, clarity, and positivity in health literacy. *Available from: https://www.igi-global. com/book/assertiveness-clarity-positivity-health-literacy/273602 (Accessed November 2021).)*

categories, there was an evident acquisition of knowledge, obtaining about 100% of correct answers. It should be noted that there were four misses in the presession, both in the professional categories of nurse and physician (Esteves & Lopes, 2021). The scientific evidence shows that effective communication is essential in the daily life of health professionals in general, since it prevents noise and misunderstandings (Silva, 2013). Nonverbal communication can complement verbal communication or replace it or contradict it, in addition to showing feelings and emotions, mainly through facial expressions. We also emphasize the importance of smiling during interactions between health professionals and patients (Schmidt, Gentry, Monin, & Courtney, 2011; Silva, 2013). If the user does not understand the message received, they will not know what to do after the consultation, which may compromise the quality and safety of the care provided. The most common cause in consultation care is attributed to failures in listening to the user (Ernesäter, Winblad, Engström, & Holmström, 2012). Another issue is not exploring key information when questioning patients (Polaschek & Polaschek, 2007). Knowing how to explore important health status issues brings safety and satisfaction to the user (Green, Spiby, Hucknall, & Foster, 2012). APTs are seen as an educational support that promotes greater learning, the application of image associated with sound provides greater memorization and motivation of the user (Ernesäter, Holmström, & Engström, 2009). Data tells us that the application of digital instruments in clinical practice is associated with great positive effects when compared to other types of intervention (Banzi et al., 2018).

The literature confirms the importance of establishing effective communication strategies in the health professional–user relationship, such as the use of simple language that facilitates the understanding of the contents addressed and adapting the information to the degree of understanding, which promotes the user's participation, in a more active way. Recognizing the way in which the service is provided in a cordial way and the importance of smiling during the interactions, providing a welcoming environment, demonstrating availability, clarifying patients' doubts, and encouraging questions are strategies that facilitate the integration of patients in their health process and, in turn, positive health outcomes (CHOP, 2015).

The third question covers "Techniques to Improve Communication" through correspondence; the trainees were asked to associate the description with the respective technique: teach-back method, follow-up, brown bag review, and educational material "handouts." In the presession, the answers revealed a lack of knowledge about the techniques, above 50% in all professional categories. However, after the intervention and exposure to this methodology, the acquisition of knowledge was notorious, approaching 100% of correct answers (Graphic 5) (Esteves & Lopes, 2021). Certain studies show that about 40% to 80% of the medical information that users receive during the consultation is immediately forgotten, and almost half of the information is incorrectly retained (CHOP, 2015). Another study highlighted the use of the message validation technique, which is still very rarely applied and has numerous advantages such as confirmation of the

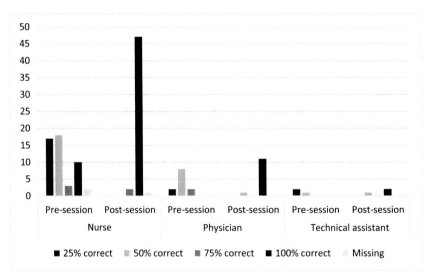

Graphic 5 Communication improvement techniques. *(Source: Esteves, S. L., & Lopes, A. S. (2021). Handbook of research on assertiveness, clarity, and positivity in health literacy. Available from: https://www.igi-global.com/book/assertiveness-clarity-positivity-health-literacy/273602 (Accessed November 2021).)*

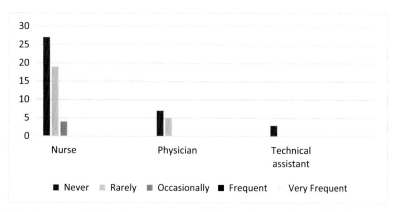

Graphic 6 Use of communication improvement techniques. *(Source: Esteves, S. L., & Lopes, A. S. (2021). Handbook of research on assertiveness, clarity, and positivity in health literacy. Available from: https://www.igi-global.com/book/assertiveness-clarity-positivity-health-literacy/273602 (Accessed November 2021).)*

information intended to be provided to the user (Ernesäter et al., 2012). A study conducted in Portugal found that about 80% of users considered a less positive experience, in the course of health professional-user care related to the lack of understanding of the information transmitted by the doctor (Gaspar, 2019). The follow-up method is another essential technique. It serves as a subsequent contact after the user's last visit. This permits the evaluation of the user's health status, the clarification of doubts, or the identification of misunderstandings that may compromise their clinical situation. The brown bag review technique should be encouraged to be used in health care settings, since it has very positive results, such as clarifying patients' doubts about the use of medication, the therapeutic indication ("what is it for"), and the dosage for "when to take it," which reduces errors (in taking, dosage, and change of medication) (CHOP, 2015). Another technique that requires improvement in its application is the delivery of educational material, such as leaflets; scientific evidence states that health professionals overdeliver this type of material, without being properly explained and analyzed together with the patient, which does not encourage interest/attention to the material (CHOP, 2015),

Finally, a question was asked about the "Use of Communication Enhancement Techniques" in the clinical practice of these health professionals, using the following scale: never, rarely, occasionally, frequently, and very frequently. According to the data obtained in Graphic 6, we found that all professional categories in this sample do not adequately apply the techniques for improving communication, since about 90% of the answers obtained fell into the categories "Never" and "Rarely," which reflects and requires intervention in this area, since health communication is a key determinant of the quality of care (Esteves & Lopes, 2021).

The scientific evidence confirms the importance of investing in training in communication between the professional and the user. Training will increase literacy, patient satisfaction, and professional safety regarding information, its quality, and treatment planning (Gaspar, 2019).

In all oral presentations given after the session, the relevance of this topic in clinical practice was questioned (if "not very relevant," "relevant," or "very relevant"). They emphasized the importance of training in health communication, namely the use of communication improvement techniques, to be addressed in health education institutions (in undergraduate courses), as well as to integrate "continuous" training in health services (Esteves & Lopes, 2021).

8 Conclusion

Health education is the combination of learning-based experiences with the goal of helping people and communities improve their health by increasing their knowledge or influencing attitudes (WHO, 2018). In this context, professionals, as health educators, must establish a helping relationship and effective communication (Carvalho & Carvalho, 2006). Effective communication is considered one of the biggest pillars for patient safety, quality of care, and the therapeutic relationship between health professionals and patients, which will benefit the integration of patients in their health process as more active and responsible agents (Direção-Geral da Saúde, 2019a, 2019b). Communication and interaction allow for improved levels of health literacy, greater autonomy, and critical reflection on choices. Indeed, this autonomy and empowerment will develop people's decision-making capacity, providing them with the means to make competent and informed decisions (Melo et al., 2009). Information should be appropriate to the literacy level of the population.

Digital health literacy, with the support and collaboration of health professionals, is considered a powerful tool in the user's health process, influencing the acquisition of higher levels of literacy, which favors access, understanding, use, and decision-making.

Lopes et al. (2020) emphasize some benefits of the application of this TAP. Some of them include:

- enhancement of learning performance and memorization in teaching and health education;
- facilitate the use of the ACP language model (Almeida, 2020);
- the integration of images associated with sound/audio is provoking emotion/retention;
- incisiveness and focus on the topic to be developed by comparison, bad practice-good practice (dichotomy);
- they are excellent means for digital dissemination (through email and social media);
- they can be replicated with other themes; and
- they are an excellent group training tool.

The results obtained with the application of this TAP, in the presession, revealed a clear lack of knowledge; about 50% of the health professionals were unaware of the techniques to improve communication. About 90% of the health professionals stated that they rarely use the techniques of improved communication in their clinical practice and pointed to a need for investment in training in health communication (Esteves & Lopes, 2021).

For health professionals to empower users to have adequate digital health literacy, it is essential that health organizations provide training and, in turn, design and implement effective, reliable, and easy-to-use digital tools for users.

References

Almeida, C. V. (2019). ACP health communication model: Communication skills at the heart of cross-cutting, holistic and practical health literacy. C. Lopes & C.V. Almeida (Coords.) In *Health literacy in practice* (pp. 43–52). Lisbon: Edições ISPA (ebook). https://repositorio.ispa.pt/bitstream/10400.12/7305/5/Literacia%20da%20Sa%C3%BAde%20na%20Pr%C3%A1tica%20%28E-.

Almeida, C. (2020). *The contribution of communication skills of doctors and nurses to health literacy: The ACP model—Assertiveness (A), Clarity (C) and Positivity (P) in the therapeutic relationship.* Available from: https://www.repository.utl.pt/handle/10400.5/20901 (Accessed March 2021).

Arroio, A., & Giordan, M. (2006). *The educational video: Aspects of teaching organization.* Available from: http://qnesc.sbq.org.br/online/qnesc24/eqm1.pdf (Accessed February 2021).

Banzi, R., Cereda, D., Moja, L., Kwag, K., Pecoraro, V., Tramacere, I., et al. (2018). *E-learning for health professionals.* https://doi.org/10.1002/14651858.CD011736.pub2 (Accessed October 2021).

Bastos, F., & Ferrari, D. (2011). Internet and patient education. *Arquivos Internacionais de Otorrinolaringologia, São Paulo-Brazil, 15*(4), 515–522.

Buntin, M., Burke, M., Hoaglin, C., & Blumenthal, D. (2011). The benefits of health information technology: A review of the recent literature shows predominantly positive results. *Health Affairs (Millwood).* Available from: http://content.healthaffairs.org/cgi/pmidlookup?view=long&pmid=21383365 (Accessed March 2021).

Carvalho, A., & Carvalho, G. (2006). *Health education: Concepts, practices and training needs.* Loures: Lusociência.

Colorado Health Outcomes Program. (2015). *AHRQ health literacy universal precautions toolkit.* Aurora: Agency for Healthcare Research and Quality.

Direção-Geral da Saúde. (2019a). *Manual of good practices health literacy. Empowering health professionals.* https://doi.org/10.13140/RG.2.2.17763.30243. Available from: https://www.dgs.pt (Accessed November 2021).

Direção-Geral da Saúde. (2019b). *Plano de Ação para a Literacia em Saúde.* Lisboa: Direção-Geral da saúde. Available from: https://www.dgs.pt (Accessed December 2021).

Directorate General of Health. (2017). *Effective communication in health care transition.* Standard number 001/2017, date 08/02/2017. Available from: https://www.dgs.pt (Accessed March 2021).

Ernesäter, A., Holmström, I. K., & Engström, M. (2009). Telenurses' experiences of working with computerized decision support: Supporting, inhibiting and quality improving. *Journal of Advanced Nursing.* Available from: http://www.ncbi.nlm.nih.gov/pubmed/19399984 (Accessed March 2021).

Ernesäter, A., Winblad, U., Engström, M., & Holmström, I. K. (2012). Malpractice claims regarding calls to Swedish telephone advice nursing: What went wrong and why? *Journal of Telemedicine and Telecare.* Available from: http://jtt.sagepub.com/content/18/7/379.long (Accessed March 2021).

Esteves, S. L., & Lopes, A. S. (2021). *Handbook of research on assertiveness, clarity, and positivity in health literacy.* Available from: https://www.igi-global.com/book/assertiveness-clarity-positivity-health-literacy/273602. (Accessed November 2021).

Eysenbach, G. (2001). What is e-health? *Journal of Medical Internet Research.* https://doi.org/10.2196/jmir.3.2.e20 (Accessed March 2021).

Free, C., Phillips, G., Watson, L., Galli, L., Felix, L., Edwards, P., et al. (2013). *The effectiveness of mobile-health technologies to improve health care service delivery processes: a systematic review and meta-analysis.* Available from: https://dx.plos.org/10.1371/journal.pmed.1001363 (Accessed March 2021).

Gaspar, T. (2019). *Gestão em Contexto Hospitalar: Modelo compreensivo da relação entre a Cultura Organizacional, fatores Psicossociais do Trabalho e Qualidade de Vida dos Profissionais de Saúde Resultados.* Centro Hospitalar Lisboa Norte da Faculdade de Medicina da Universidade de Lisboa.

Green, J., Spiby, H., Hucknall, C., & Foster, H. (2012). Converting policy into care: Women's satisfaction with the early labour telephone component of the All Wales Clinical Pathway for Normal Labour. *Journal of Advanced Nursing.* Available from: http://onlinelibrary.wiley.com/doi/10.1111/j.1365-2648.2011.05906.x/abstract (Accessed March 2021).

Gÿörffy, Z., & Mesko, B. (2019). *The rise of the empowered physician in the digital health era: Viewpoint.* Available from: https://www.jmir.org/2019/3/e12490/ (Acessed October 2021).

Infopédia. (2021). https://www.infopedia.pt/dicionarios/lingua-portuguesa/audiovisual (Accessed October 2021).

Latulippe, K., Hamel, C., & Giroux, D. (2017). Social health inequalities and eHealth: A literature review with qualitative synthesis of theoretical and empirical studies. *Journal of Medical Internet Research, 19*(4), e136.

Lopes, A. S., Filipe, B., & Esteves, S. L. (2020). Segurança no Comunicar—Uma técnica audiovisual pedagógica. C. V. Almeida, K. Moraes & V. V. Brasil (Coords.) In *50 Técnicas de literacia em saúde na prática. Um guia para a saúde* (pp. 110–112). Alemanha: Novas Edições Académicas.

Mackert, M., Kahlor, L., Tyler, D., & Gustafson, J. (2009). Designing e-health interventions for low-health-literate culturally diverse parents: Addressing the obesity epidemic. *Telemedicine Journal and E-Health.* Available from: http://europepmc.org/abstract/MED/19694596 (Accessed March 2021).

Mayer, R. E. (2014). *The Cambridge handbook of multimedia learning.* New York: Cambridge University Press. Available from: https://assets.cambridge.org/97811070/35201/frontmatter/9781107035201_frontmatter.pdf (Accessed February 2021).

Meirelles, A., Sena, M., Cruz, I., Luedy, A., & Junior, H. (2013). The role of patient and family education in building a process of safety and quality in a University Hospital. *Revista Acreditação, 3*(8), 23–33.

Melo, M., et al. (2009). A educação em saúde como agente promotor de qualidade de vida para o idoso. *Ciência Saúde Coletiva, 14*(1), 1579–1586.

Mendonça, L. G., Ferreira, F. R., & Rodriguez, L. L. R. (2014). *Production as a didactic resource for teaching legislation in an undergraduate course in Chemistry. Química Nova na Escola.* Available from: 06-RSA-135-12.pdf (http://sbq.org.br). (Accessed February 2021).

Ministry of Health. (2013). *Labour and Welfare Guidance for implementation of specific health guidance utilizing ICT.* Available from: https://www.mhlw.go.jp/bunya/shakaihosho/iryouseido01/dl/info03j-130822_04.pdf (Accessed March 2021).

National Health Service. (2020). *Chunk & Check.* The Health Literacy Place. Available from: http://www.healthliteracyplace.org.uk/tools-and-techniques/techniques/chunk-and-check/ (Accessed March 2021).

Nijland, N., van Gemert-Pijnen, J., Boer, H., Steehouder, M. F., & Seydel, E. R. (2018). Evaluation of Internet-based technology for supporting self-care: Problems encountered by patients and caregivers when using self-care applications. *Journal of Medical Internet Research.* https://doi.org/10.2196/jmir.957 (Accessed October 2021).

Norgaard, O., Furstrand, D., Klokker, L., Karnoe, A., Batterham, R., Kayser, L., et al. (2015). The e-health literacy framework: A conceptual framework for characterizing e-health users and their interaction with e-health systems. *Knowledge Management & E-Learning, 7*(4), 522–540.

Norman, C., & Skinner, A. (2006). eHealth literacy: Essential skills for consumer health in a networked world. *Journal of Medical Internet Research.* Available from http://www.jmir.org (Accessed March 2021).

Polaschek, L., & Polaschek, N. (2007). Solution-focused conversations: A new therapeutic strategy in well child health nursing telephone consultations. *Journal of Advanced Nursing.* Available from: http://onlinelibrary.wiley.com/doi/10.1111/j.1365-2648.2007.04314.x/abstract (Accessed March 2021).

Rankin, S. (2001). *Patient education: Principles and practice* (4th ed.). Philadelphia: Lippincott Williams & Wilkins.

Reed, K. (2010). *Therapeutic patient education.* Disponível em: www.diabetesinfo.org.nz/librarary/pe_therapeutic.pdf (Acedido em outubro 2021).

Schmidt, K. L., Gentry, A., Monin, J. K., & Courtney, K. L. (2011). Demonstration of facial communication of emotion through telehospice videophone contact. *Telemedicine Journal and E-Health*. Available from: http://online.liebertpub.com/doi/abs/10.1089/tmj.2010.0190 (Accessed March 2021).

Scott, M. D. (2019). Best practices in digital health literacy. *International Journal of Cardiology*. https://doi.org/10.1016/j.ijcard.2019.05.070 (Accessed March 2021).

Silva, M. J. P. (2013). *Comunicação tem remédio: A comunicação nas relações interpessoais em saúde* (9th ed.). São Paulo: Editora Loyola.

Sociedade Portuguesa de Diabetologia. (2018). *Therapeutic education in diabetes – Competencies of health professionals and people with diabetes-2018*. SPD. http://spd.pt/images/booklet_educação_terapeutica.pdf.

Sorensen, K., Broucke, S. V., Fullam, J., Doyle, G., Pelikan, J., Slonska, Z., et al. (2012). *Health literacy and public health: A systematic review and integration of definitions and models*. BMC Public Health.

Tang, P. C., Ash, J. S., Bates, D. W., Overhage, J. M., & Sands, D. Z. (2006). Personal health records: Definitions, benefits, and strategies for overcoming barriers to adoption. *Journal of the American Medical Informatics Association, 13*(2), 121–126.

Thomas, J., Barraket, J., Wilson, C. K., Rennie, E., Ewing, S., & MacDonald, T. (2019). *Measuring Australia's digital divide: The Australian digital inclusion index 2019*. Melbourne: RMIT University and Swinburne University of Technology, for Telstra.

U.S. Department of Health and Human Services. (2000). *Healthy people 2010*. Washington: U.S. Government Printing Office.

U.S. Department of Health and Human Services. (2015). *Health information technology*. Washington, DC: Office for Civil Rights. Available from: http://www.hhs.gov/hipaa/for-professionals/special-topics/health-information-technology/index.html (Accessed March 2021).

World Health Organization. (1998). *Health promotion glossary*. Geneva: WHO. Available from: https://www.who.int/publications/i/item/WHO-HPR-HEP-98.1 (Accessed August 2021).

World Health Organization. (2006). *Education at a glance*. https://doi.org/10.1787/19991487 (Accessed September 2021).

World Health Organization. (2018). *Global Action Plan for Physical Activity 2018–2030 More active people for a healthy world*. Available from: https://apps.who.int/iris/bitstream/handle/10665/272721/WHO-NMH-PND-18.5-por.pdf (Accessed September 2021).

World Health Organization. (2020). *Health literacy toolkit for low- and middle-income countries*. Available from: http://www.searo.who.int/entity/healthpromotion/documents/hl_tookit/en/ (Accessed March 2021).

CHAPTER 4

The HOPE project—A case study on the development of a serious game to increase pediatric cancer patients' motivation

Hernâni Zão Oliveira[a], Nuno Patraquim[b], and Helena Lima[b]
[a]Department of Management and Innovation, Colégio do Espírito Santo, University of Évora, Évora, Portugal
[b]University of Porto, Porto, Portugal

1 Introduction

The level of knowledge a patient has about cancer can influence treatment adherence (Davis, Williams, Marin, Parker, & Glass, 2002). Effectively, the transfer of information and knowledge about the disease and its treatment can change the individual's dysfunctional ideas, which consequently improves their ability to cope with the disease and, ultimately, encourages adherence (van Dulmen et al., 2007). Bryan, Kelly, Chesters, et al. (2021) state that children have a lack of knowledge about Biology topics until the age of 10, which means that they have difficulties understanding cancer. Studies show that, at this age range, children believe that people get sick if they come into contact with other people and with objects that have bad properties or as a form of punishment for having performed a bad deed (Bares & Gelman, 2008). Knighting, Rowa-Dewar, Malcolm, Kearney, and Gibson (2011) showed that children view cancer in a negative way from an early age, even without personal experience.

As a result of the evolution of treatments in pediatric oncology, there has been a significant increase in survival rates in children with cancer and, consequently, a constant growth in the population of survivors (Siegel, Naishadham, & Jemal, 2013). Despite these developments, pediatric cancer remains associated with a wide spectrum of disease-related side effects, which can progress to chronic consequences and result in a long-term negative impact. To this panorama is added, very often, the impact on the social, psychological, and physiological levels of the surviving individual (Florin et al., 2007).

Inactivity, impairment of cardiopulmonary and musculoskeletal function, and reduced levels of motor performance (De Souza et al., 2007) and cognitive abilities (Krull et al., 2011) are frequently detected. Current studies also show a negative impact

Active Learning for Digital Transformation in Healthcare Education, Training and Research
https://doi.org/10.1016/B978-0-443-15248-1.00001-1

on psychological well-being, satisfaction, and social functioning, which together determine a marked decrease in the survivor's quality of life (Wright, Galea, & Barr, 2005).

Over the last few years, several studies have demonstrated the first scientifically valid evidence of the positive effects of interventions that incorporated physical exercise as an integral part of the cancer treatment routine (Wampler et al., 2012). First results show an association between increased levels of physical activity in cancer patients and an improvement in quality of life during hospitalization (San Juan, Wolin, & Lucía, 2011). In particular, physical functioning is increased, anxiety is reduced, and social integration is encouraged (San Juan, Chamorro-Viña, Moral, et al., 2008). Considering the fact that physical activity plays a vital role in children's physiological and psychosocial development, therapeutic exercise in pediatric oncology is particularly important to create a better bodily response to treatments (Haemmerli, Ammann, Roessler, Koenig, & Brack, 2022). However, there is still a lack of comprehensive and evidence-based data in the field of interventions that incorporate physical exercise in pediatric wards, given the constraints that this type of intervention entails.

1.1 Serious games: A new opportunity for health education

Many psychoeducational interventions have already been developed to change the patient's attitude and knowledge regarding their treatment (Baranowski, Buday, Thompson, & Baranowski, 2008). Nonetheless, reviews on the topic indicate that these vary widely in their effectiveness. This randomness becomes particularly evident in adolescent and young adult patients (Kliem & Wiemeyer, 2010).

It has been suggested that this type of intervention could be more effective by benefiting from the use of multimedia resources, such as video games, as a preferred recreational activity for this target audience (Papastergiou, 2009). Serious games are being highlighted on this matter, as video games with the aim of transmitting educational or training content to the user. These resources simulate real-world situations or processes that are designed to solve a problem and have been shown to be particularly relevant in the development of new therapies and health tools (Kato, Cole, Bradlyn, & Pollock, 2008).

Serious games have potential advantages as vehicles for learning and persuasion, including high accessibility, easy content updating, low cost per person, high interactivity, and high individualization of use, with the ability to incorporate attractive graphics, including animation and virtual reality (Papastergiou, 2009).

Evidence has become consistent in reporting the impact of using these resources in psychoeducational interventions for adolescents and young adults with chronic illnesses. A particular study compared the effect of application of a video game on the self-management of asthma in children aged 9 to 13 years with that on a control group that did not receive any intervention. It was found here that children who received the game had a higher level of knowledge about self-regulation, treatment, and prevention, even after adjusting for differences between groups in pre-intervention knowledge levels. About 84% of the intervention group chose to continue using the game after completing the study (Kliem & Wiemeyer, 2010).

In the field of oncology, another study compared two small groups of children (4–11 years old) with leukemia who received two different types of intervention: a video game (Kidz with Leukemia) or a leaflet informative. The results showed an increase in the impact of the video game when compared to the information leaflet, ensuring greater motivation of the target audience (Kretschmann, Dittus, Lutz, & Meier, 2010).

1.2 Serious games, physical activity, and intrinsic motivation

Physical exercise programs have provided improvements in quality of life and reduction of fatigue in cancer patients (Haemmerli et al., 2022). The impact of physical exercise has also been shown to increase aerobic capacity, improving vitality and emotional well-being and decreasing depression states (Huang, 2011). The practice of physical exercise motivated by recreational resources such as video games, called exergames, has been associated with greater pleasure in performing activities, facilitating adherence to therapy and rehabilitation processes (Benzing & Schmidt, 2018).

Studies on the impact of this type of technology have been particularly evident in the reduction of fatigue in different stages of the disease and treatment in cancer patients, with improvements evident in the motor conditions, with reinforcement of the muscular condition (San Juan et al., 2011). Clinical trials have been particularly relevant in demonstrating the effectiveness of physical exercise in the hospitalization context for emotional motivation and the creation of more stable physical conditions (Knols, Aaronson, Uebelhart, Fransen, & Aufdemkampe, 2005).

According to Prensky (2006), there are several elements that influence involvement, but fun and pleasure/satisfaction are the ones that have the greatest impact. Veletsianos and Doering (2010) indicate that these two variables support and reinforce learning and are useful elements in children's cognitive development. Several studies recommend an increase in the pleasure/satisfaction perceived by children, as this increase will result in greater intrinsic motivation. Stimulating intrinsic motivation is important to increase the attractiveness of serious games and is widely accepted as a desirable educational practice, because it leads to long-term motivation and continued participation in activities (Karimi & Lim, 2010).

There are authors who recommend the direct observation of different behaviors, such as changes in facial expressions, as a method of evaluating engagement and motivation (Read, MacFarlane, & Casey, 2001). But when the target audience is children, this direct observation assessment is often subjective (Karimi & Lim, 2010). Intrinsic motivation scales are instruments commonly used to understand the degree of involvement of children in video games (Read & MacFarlane, 2000).

1.3 The HOPE project

The conceptualization of the narrative of The Hope Project, a video game for children hospitalized with cancer, followed an early co-design approach, derived from co-operative inquiry (Druin, 1999). This strategy of working with children combines and adapts the low-tech prototyping of participatory, observation, and note-taking techniques of contextual inquiry and the time and resources of technology immersion (Galler, Kristine, Myhrer, & Varela, 2022).

Narrative plays an important role in the user experience. It increases motivational levels in serious games and has positive effects on learning (Mendonça & Mustaro, 2011). The basis of the narrative to be validated for this video game followed a three-act structure.

Act 1 - Beginning: The hero feels a strong stomachache and goes to the hospital. At the doctor's office, the hero is observed. Due to the doctor's many doubts, the character is advised to undergo diagnostic tests. After carrying out the tests, the hero is diagnosed with cancer.

Act 2 - Intermediate: The hero is hospitalized and the fight against cancer begins. There are several challenges that the character will have to face, from different treatments to choices of food and exercise. In the course of the various challenges, the character will have powers and allies that will help to fight the disease.

Act 3 - End: The character overcomes the various challenges. The hero is congratulated on his achievements. Before leaving the hospital, the most important information is reviewed in order to be respected on a daily basis.

Game mechanics can be defined as "rules based on systems/simulations that facilitate and encourage the player to explore and learn the properties of their space of possibilities, through the use of feedback mechanisms" (Cook, 2006). However, in the development of video games for educational purposes, it is necessary to consider the learning mechanics that work as principles for the construction of game mechanics. Learning mechanics indicate which rules should be applied but do not describe the corresponding game mechanics (El-Nasr, Drachen, & Canossa, 2013). In the prototype of this project, the mechanics of learning go through explaining what an oncological disease is and what tests are carried out to detect it.

Thus, the game mechanics of the different challenges of the HOPE Project were the following:

Blood Analysis: In this challenge, the player has to hit the target with a "virtual" syringe on the arm of the character who is shaking, using the touch on the screen. Although the syringe is present, the child does not control it. The user just tries to hit the target with clicks on the character. The child has 10 attempts to hit the target.

Urine Analysis: In this challenge, the user has to catch the drops of urine that fall from the top edge of the screen. The user has to control a cup with the touch and drag of the finger, which only moves horizontally. This challenge has a timer that decreases as the challenge progresses.

X-ray: The goal is to perform certain corporal poses for a period of time. The poses are captured and analyzed through the front camera of mobile devices. There are three different poses. For each pose, the user has to hold for 5 consecutive seconds to perform it. This will help to strengthen the child's muscles. In addition, they have three execution attempts for each pose.

Biopsy: the objective is to control a syringe by touching and dragging the finger through the skin to an inflamed lymph node. After reaching the ganglion cyst and taking a sample, the child is asked to lead the syringe out of the character's body. There is a path drawn on the skin that the child must respect when using the syringe. In addition, there is also a timer. If the players touch outside the limits of the dashed line, or if they don't collect the sample in the desired time, they have to restart the challenge.

Cell Proliferation: the goal is to destroy as many cancer cells as possible. A particularity of this challenge is that the child will never be able to destroy all the cancer cells and the challenge ends when a cancer cell reaches the edge of the screen (it serves as an introduction to the theme of the next challenge - metastasis). Despite this, the final score is reflected in the number of cancerous and non-cancerous cells that the child has destroyed. To destroy the cells, the player has to click on them.

Metastasis: In this challenge, the goal is to command a spaceship, which travels through the character's blood vessels to destroy the largest number of cancer cells (which are cells that escaped from the previous challenge), thus preventing metastasis. The spaceship is controlled by the child's movements. If the child walks sideways, the spacecraft follows him; if the user performs a squat, the spacecraft descends and goes up again; and if the player jumps, the spacecraft goes up and down again. The child must destroy the cancer cells, guiding the spaceship to their position and avoiding obstacles. If the users do not destroy any cancer cells or if they collide with obstacles three times, they have to start the challenge again.

2 Methodology

To better develop The HOPE Project, a research work was conducted based on two specific objectives: (1) to understand the routines of children in relation to video games and (2) to co-create specific parts of the narrative with children hospitalized with cancer.

2.1 Understanding the routines of children in relation to video games

The first part of the research was achieved with the implementation of a characterization survey to children between 6 and 12 years old. Due to their cognitive immaturity, it was necessary to apply some principles for the construction of the instrument (De Leeuw, 2011), namely, avoiding ambiguity in questions and scales, indirect questions, the construction of complex questions (the use of period and comma, colon, parentheses, or subordinate sentences) and avoiding asking questions in the negative form and questions with numerical quantities.

For the analysis of the differences and associations between the variables, the chi-square and Cramer's V tests were used, analyzing the value of statistical significance through the P value equal to or less than 0.05.

2.2 Co-creating specific parts of the narrative with children hospitalized with cancer

The co-creation sessions were developed with children hospitalized with cancer in two important meetings.

Meeting #1: An activity was developed in order to obtain feedback about the protagonist. This activity was essentially divided into three tasks:

1st task: Drawing of a superhero and their respective powers. The only restriction imposed was that the character had to be bald.

2nd task: Participation in a memory game, using facial expressions of a prototype character.

3rd task: Participation in an emotional recognition exercise, using the facial expressions of the second task.

In the second co-creation session, the participants were able to experience a prototype of The HOPE Project serious game, and their intrinsic motivation was measured. For this measurement, a subscale of the IMIp (Portuguese version of Intrinsic Motivation Inventory) was chosen.

The Intrinsic Motivation Inventory (IMIp) is a translated version and was adapted by Fonseca (1995) (McAuley, Duncan, & Tammen, 1989). This instrument has become quite popular because it allows the assessment of the intrinsic motivation of individuals in relation to any activity, since the items allow a specific adaptation to the context/activity of the participants. Furthermore, its global psychometric properties remain stable regardless of the use of only some of its subscales or the reduction in the number of items per subscale (Fonseca & Brito, 2001). This questionnaire, in the Portuguese version, consists of four subscales—Pleasure/Interest, Competence, Effort/Importance, and Tension/Pressure—but only Pleasure/Interest effectively assesses the intrinsic motivation of participants in a given activity (Karimi & Lim, 2010). The questionnaire developed for this research work contains the Pleasure/Interest subscale adapted to the activity of playing video games.

The participants were asked to indicate their degree of agreement with each statement, on a scale of 1 (Strongly disagree) to 5 (Strongly agree). For the construction of the scale, the Smileyometer was used. This response method is advised for children when the number of choice options is greater than 3 (De Leeuw, 2011).

3 Results

3.1 Characterization survey

In total, 78 responses, 40 from female and 38 from male participants, were obtained. The average age of the participants was 10.22 (SD = 1.593), with a range of 6 to 12. Upon asking the question, "Do you like to play video games?", it was found that 71.8% of respondents really like to play, while 25.6% like it moderately. In terms of cumulative percentage, 97.4% of children really and moderately enjoy playing. Only 2.6% of the surveyed population (two children) responded that they do not enjoy playing video games.

Although there was no statistically significant association between the variable "enjoy playing video games" and the variable "enjoy playing on tablets" (V = 0.249, P = 0.063), it was found that a large percentage of children who play video games in general also enjoy playing on tablets (73.6%). Still, in this group, 22.6% moderately like to play games on these devices, which gives a cumulative percentage of 96.2%. Around 88% of the children who moderately enjoy playing video games like to play on tablets. The rest of these (11.8%) like to play on tablets "moderately". That is, in this group, none of the children responded that they did not like to play games on this mobile device (Fig. 1).

According to the results (Fig. 2), children who like to play video games on tablets are more likely to have their own device (59.3%). There is also a tendency to have their own device in the group of children who like to play moderately (50%). However, there is a greater tendency for children who like to play moderately to use the devices of family

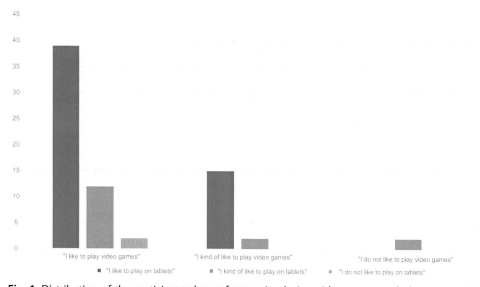

Fig. 1 Distribution of the participants by preference in playing video games and playing on tablets.

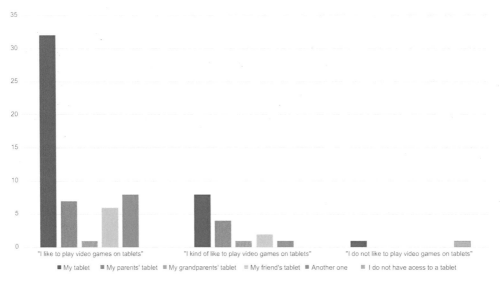

Fig. 2 Distribution of the participants who play video games by the answers given to the following questions: "Do you like playing video games on tablets?" and "What tablet do you usually play on?"

members (parents and grandparents) (31.3%) than for children who really like to play (15%). Only the group of children who do not like to play video games on tablets answered that they do not have access to these devices (50%).

There are significant differences between genders in terms of their taste in playing video games ($\chi = 8740$, $P = 0.013$), since there are more boys who like to play video games (86.8%) than girls (57.5%). There is an association between the gender and the time spent per week with video games ($V = 0.438$, $P = 0.000$).

3.2 Co-creation sessions to obtain feedback about the video game's protagonist

Thirteen children hospitalized with cancer participated in the co-creative sessions, and each of them drew a possible hero for The HOPE Project video game (Fig. 3).

In addition to developing a character with whom the children can identify, these activities were important to understand whether the character's expressions were perceptible to the participants. Through a memory game (Fig. 4), it was detected that children had difficulties identifying differences between expressions, namely cards (d) and (h).

The emotional recognition exercise was an important challenge, as expressions that needed to be modified were detected, namely, facial expressions in which no one could identify an underlying state of mind (card [g]); facial expressions in which children identified different emotions (cards [c], [d], and [f]); and facial expressions in which the mood identified by the children was not similar to the mood previously assigned by the development team (card [h]).

Fig. 3 Some of the drawings made by the participants during the co-creation sessions.

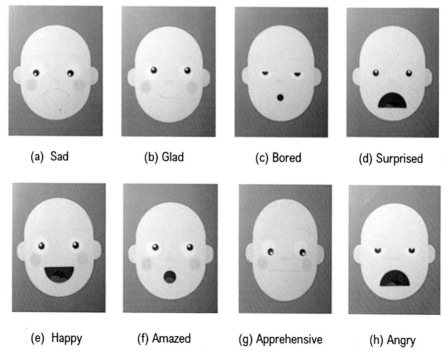

(a) Sad   (b) Glad   (c) Bored   (d) Surprised

(e) Happy   (f) Amazed   (g) Apprehensive   (h) Angry

Fig. 4 List of the cards used in the expression identification exercise.

Thus, on card (c), states of mind were interpreted differently, such as: "You are sleepy", "You are sad", "You are amazed," or "You are annoyed". On card (d), the following were marked: "You are amazed", "You are screaming," and "He is asking for help". And on card (f), the following observations were made: "He is crying", "He is angry", "He is impressed," and "He is surprised". In card (h), the children assumed that the character was annoyed, when the development team had assigned the angry state to the card.

Finally, when the narrative of the serious game was shared with the audience, three types of characters were defined in collaboration with the participants:

The hero character—the player's avatar. It is the character who has cancer, and who seeks to fight it, through the knowledge that is acquired. The hero starts out as a child like so many others, but when it is diagnosed with an oncological disease, it becomes a super-hero with various powers. Many superheroes hide their identities through disguises, and this one is no exception. The superhero uses its pajamas as a combat uniform.

The mentor character: It is the character that guides the hero. In this video game, the mentor is the figure of a medical doctor who helps the hero with decision-making strategies throughout the game.

The allies characters: These are the characters that accompany the hero in different stages of the game. In this serious game, the allies are family members (father and mother), friends, and other people, such as the nurse, the physiotherapist, the educator, and the volunteer.

3.3 Intrinsic motivation evaluation

The internal consistency of the Pleasure/Interest dimension evaluated in this study was satisfactory ($\alpha = 0.741$).

The results presented illustrate a deep satisfaction and pleasure of children upon their first session playing with this video game (Fig. 5). No one responded below 4 on the Likert scale available, which means that the children either agreed or fully agreed with the statements. And even with an answer range between 4 and 5, it is also verified that, in all questions, more than 50% of the answers are answered with the highest value ("I totally agree").

Besides this questionnaire, direct observation of the children's behavior also allowed drawing conclusions about their motivation and involvement. Laughs, celebrations when they successfully completed the challenges, and the perceived attention with which they played were elements that showed the positive effect of the activity on these children.

In this regard, some of the phrases spontaneously uttered by the participants during the game session were pointed out: "When I get home, can I download it?"; "This video game is really cool."; "The nurse is waiting for me, but I will continue to play."; "I have an Android tablet. Can I play there?"

Pleasure/Interest	1	2	3	4	5	Mean	SD
Q1. I had a lot of fun playing this video game	0	0	0	1	12	4.92	0.277
Q2. I really liked this video game	0	0	0	3	10	4.77	0.439
Q3. This video game is very interesting	0	0	0	2	11	4.85	0.376
Q4. While I was playing I was enjoying it	0	0	0	3	10	4.77	0.439
Q5. This video game caught my attention	0	0	0	6	7	4.54	0.519

Fig. 5 Results of the intrinsic motivation tests, using the IMIp's subscale "Pleasure/Interest".

4 Discussion

The conception and design of a video game is a complex process. It involves the simultaneous coordination of several phases: the conceptualization of the theme; the creation of characters; the construction of the narrative; the definition of game mechanics; the design of the levels; and the definition of interfaces, among others. This complexity is increased when, in addition to entertaining, they aim to promote knowledge and have an impact on health.

Techniques combined with gamification involve and empower the patient, make the recovery process easier and less expensive, and can also ensure more sustainable monitoring by health professionals. The application of these types of solutions can bring competitiveness to companies in the health sector at a regional level, with a direct impact on society and with a direct impact on society and the well-being of the individual.

For this reason, it is very important, whenever possible, to integrate the target audience into the process of developing these solutions. This project sought to know in more detail children's routine, in order to understand the relevance of creating a video game and the type of platform it should run on.

Children are motivated to play these types of games either because it uses familiar platforms, they like interaction with the graphics created, or the introduction of new mechanics captures their attention, such as exergaming. In fact, in this research protocol, no child had ever tried a video game for tablets with this component.

However, the results also showed that more than half of the constructed facial expressions required modifications, which justifies the importance of including children in the various stages of video game construction. This will be crucial for the development of a tool to increase knowledge and enhancement of behavior change.

The HOPE Project was designed to unite the realistic perspective of cancer with a fantastic world. The realistic perspective is achieved through the illustrations used in the video game that are based on photo reports from hospital spaces. With this, children can recognize on the platform all the reality they go through, being prepared to face new clinical procedures. The fantastic world is introduced to them with the narrative that tells the story of a child, the main character, who fights cancer like the users are fighting too.

The introduction of this analogy becomes important to promote the sick child's motivation and understanding of what causes him discomfort. Through entertainment, this project intends to shape the child's behavior to make hospital routines more effective, with a reduction of time to stabilize the patient for medical procedures. The video game uses interactive processes to explain, for example, the importance of the child remaining calm when taking a blood sample and implements movement, coordination, and elasticity challenges to explain what an X-ray is.

The hair loss is also perceived as a side effect of the treatment, but it is associated with a new superpower of the superhero, along with his pajamas. The symbolism of associating a hairless child with a superhero tries to counteract the children's negative image of their visual appearance.

The involvement of children in this research work was very important. Still, the number of co-creation sessions was very short. It is possible to conclude that this video game had a very positive initial impact on children, but it would take more game sessions spread over time to effectively conclude that the novelty effect is not the main cause of these results. Because of this, future work in this area should address this issue.

5 Conclusions

The HOPE Project was born with the purpose of finding solutions to certain difficulties identified in children with cancer who are in a clinical context.

Indeed, there seem to be constant obstacles encountered in the willingness of children to undergo treatments, and it is also verified that they have little knowledge about the disease and have a sedentary lifestyle.

The results presented in this study demonstrate that the video game Hope is a tool with the potential to promote health education and the motivation of its users. This motivation can translate into an increase in information related to strategies to overcome debilitating problems.

Despite having the hospital as the initial setting, it is intended to encourage the user to return to active life again, which is why this project will include a school setting and a home setting with levels that will playfully work the main challenges felt in these contexts.

It is hoped that, with this case study, the reader can reflect on the following points:

1. Promoting education in a hospital context is very important, and the use of innovative platforms should be encouraged;
2. The development of new solutions must include a previous period in order to analyze in detail the routines of the target audience and their main needs;
3. The conceptualization and validation of solutions is reinforced by the integration of the target audience in co-creation sessions;
4. The HOPE Project is just one example of the potential of new technologies that can involve the patient, making patient journeys more manageable.

Readers interested in this topic are encouraged to consult the following projects about health and co-creation:

- HEALTH CASCADE: https://healthcascade.eu/health-cascade-evidence-based-co-creation-for-public-health/
- HCARE: https://hcarecanada.com/co-creation/
- CODE FOR HEALTH: https://codeforhealth.de/en/projects/cocrehit/

References

Baranowski, T., Buday, R., Thompson, D. I., & Baranowski, J. (2008). Playing for real. Video games and stories for health-related behavior change. *American Journal of Preventive Medicine*, *34*(1), 74–82. https://doi.org/10.1016/j.amepre.2007.09.027.

Bares, C. B., & Gelman, S. A. (2008). Knowledge of illness during childhood: Making distinctions between cancer and colds. *International Journal of Behavioral Development*, *32*(5), 443–450. https://doi.org/10.1177/0165025408093663.

Benzing, V., & Schmidt, M. (2018). Exergaming for children and adolescents: Strengths, weaknesses, opportunities and threats. *Journal of Clinical Medicine*, 7, 422. https://doi.org/10.3390/jcm7110422.

Bryan, G., Kelly, P., Chesters, H., et al. (2021). Access to and experience of education for children and adolescents with cancer: A scoping review protocol. *Systematic Reviews*, *10*, 167. https://doi.org/10.1186/s13643-021-01723-4.

Cook, D. (2006). *What are game mechanics?*. lostgarden.com. Available from: http://lostgarden.com/2006/10/what-are-game-mechanics.html.

Davis, T. C., Williams, M. V., Marin, F., Parker, R. M., & Glass, J. (2002). Health literacy and cancer communication. *CA: a Cancer Journal for Clinicians*, *52*, 134–149. https://doi.org/10.3322/canjclin.52.3.134.

De Leeuw, E. D. (2011). *Improving data quality when surveying children and adolescents: Cognitive and social development and its role in questionnaire construction and pretesting*. Available from: https://www.aka.fi/globalassets/tietysti1.fi/awanhat/documents/tiedostot/lapset/presentations-of-the-annual-seminar-10-12-may-2011/surveying-children-and- adolescents_de-leeuw.pdf.

De Souza, A. M., Potts, J. E., Potts, M. T., De Souza, E. S., Rowland, T. W., Pritchard, S. L., et al. (2007). A stress echocardiography study of cardiac function during progressive exercise in pediatric oncology patients treated with anthracyclines. *Pediatric Blood & Cancer*, *49*(1), 56–64. https://doi.org/10.1002/pbc.21122.

Druin, A. (1999). Cooperative inquiry: Developing new technologies for children with children. In *Proceedings of the ACM SIGCHI 99 conference on human factors in computing systems, Pittsburgh, USA* (pp. 592–599).

El-Nasr, S. M., Drachen, A., & Canossa, A. (2013). In M. S. El-Nasr, A. Drachen, & A. Canossa (Eds.), *Game analytics, maximizing the value of player data* Springer.

Florin, T. A., Fryer, G. E., Miyoshi, T., Weitzman, M., Mertens, A., Hudson, M., et al. (2007). Physical inactivity in adult survivors of childhood acute lymphoblastic leukemia: A report from the childhood cancer survivor study. *Cancer Epidemiology, Biomarkers & Prevention*, *16*, 1356–1363. https://doi.org/10.1158/1055-9965.EPI-07-0048.

Fonseca, A. M. (1995). *Versão portuguesa do Intrinsic Motivation Inventory (IMI): Inventário de Motivação intrínseca (IMIp).*

Fonseca, A. M., & Brito, A. P. (2001). Propriedades psicométricas da versão portuguesa do Intrinsic Motivation Inventory (IMIp) em contextos de actividade física e desportiva. *Análise Psicológica*, 59–76.

Galler, M., Kristine, S., Myhrer, G. A., & Varela, P. (2022). Listening to children voices in early stages of new product development through co-creation – Creative focus group and online platform. *Food Research International*, *154*, 111000. https://doi.org/10.1016/j.foodres.2022.111000.

Haemmerli, M., Ammann, R. A., Roessler, J., Koenig, C., & Brack, E. (2022). Vital signs in pediatric oncology patients assessed by continuous recording with a wearable device, NCT04134429. *Scientific Data*, *9*(1). https://doi.org/10.1038/s41597-022-01182-z.

Huang, W.-H. (2011). Evaluating learners' motivational and cognitive processing in an online game-based learning environment. *Computers in Human Behavior*, *27*(2), 694–704.

Karimi, A., & Lim, Y. P. (2010). Children, engagement and enjoyment in digital narrative. In *Proceedings of ASCILITE - Australian society for computers in learning in tertiary education annual conference 2010* (pp. 475–483).

Kato, P. M., Cole, S. W., Bradlyn, A. S., & Pollock, B. H. (2008). A video game improves behavioural outcomes in adolescents and young adults with cancer: A randomized trial. *Pediatrics*, *122*(2), 305–317. https://doi.org/10.1542/peds.2007-3134.

Kliem, A., & Wiemeyer, A. (2010). Comparison of a traditional and a video game based balance training program. In *Proceedings of the GameDays 2010 – Serious games for sports and health, Darmstadt* (pp. 37–50).

Knighting, K., Rowa-Dewar, N., Malcolm, C., Kearney, N., & Gibson, F. (2011). Children's understanding of cancer and views on health-related behaviour: A 'draw and write' study. *Child: Care, Health and Development*, *37*(2), 289–299. https://doi.org/10.1111/j.1365-2214.2010.01138.x.

Knols, R., Aaronson, N. K., Uebelhart, D., Fransen, J., & Aufdemkampe, G. (2005). Physical exercise in cancer patients during and after medical treatment: A systematic review of randomized and controlled clinical trials. *Journal of Clinical Oncology*, *23*, 3830–3842. https://doi.org/10.1200/JCO.2005.02.148.

Kretschmann, R., Dittus, I., Lutz, I., & Meier, C. (2010). Nintendo Wii sports: Simple gadget or serious "measure" for health promotion? – A pilot study according to the energy expenditure, movement extent, and student perceptions. In *Proceedings of the GameDays 2010 – Serious games for sports and health, Darmstadt, 2010* (pp. 147–159).

Krull, K. R., Annett, R. D., Pan, Z., Ness, K. K., Nathan, P. C., Srivastava, D. K., et al. (2011). Neurocognitive functioning and health-related behaviours in adult survivors of childhood cancer: A report from the childhood cancer survivor study. *European Journal of Cancer*, *47*(9), 1380–1388. https://doi.org/10.1016/j.ejca.2011.03.001.

McAuley, E., Duncan, T., & Tammen, V. V. (1989). Psychometric properties of the intrinsic motivation inventory in a competitive sport setting: A confirmatory factor analysis. *Research Quarterly for Exercise and Sport*, *60*(1), 48–58. https://doi.org/10.1080/02701367.1989.10607413.

Mendonça, R. L., & Mustaro, P. N. (2011). Como tornar aplicações de realidade virtual e aumentada, ambientes virtuais e sistemas de realidade mista mais imersivos. In *Symposium on virtual and augmented reality, Uberlândia/MG. Anais. Uberlândia/MG: Editora SBC.*

Papastergiou, M. (2009). Exploring the potential of computer and video games for health and physical education: A literature review. *Computers & Education*, *53*(3), 603–622. https://doi.org/10.1016/j.compedu.2009.04.001.

Prensky, M. (2006). *Don't bother me, Mom—I'm learning. How computer and video games are preparing your kids for 21st century success—And how you can help!.* Saint Paul, MN: Paragon House.

Read, J. C., & MacFarlane, S. J. (2000). Measuring Fun. In *Computers and fun 3*. York: England.

Read, J. C., MacFarlane, S. J., & Casey, C. (2001). Expectations and endurability - measuring fun. In *Computers and fun 4*. York: England.

San Juan, A. F., Chamorro-Viña, C., Moral, S., et al. (2008). Benefits of intrahospital exercise training after pediatric bone marrow transplantation. *International Journal of Sports Medicine, 29*, 439–446. https://doi.org/10.1055/s-2007-965571.

San Juan, A. F., Wolin, K., & Lucía, A. (2011). Physical activity and pediatric cancer survivorship. *Recent Results in Cancer Research, 186*, 319–347. https://doi.org/10.1007/978-3-642-04231-7_14.

Siegel, R., Naishadham, D., & Jemal, A. (2013). Cancer statistics, 2013. *CA: a Cancer Journal for Clinicians, 63*, 11–30.

van Dulmen, S., Sluijs, E., van Dijk, L., de Ridder, D., Heerdink, R., & Bensing, J. (2007). Patient adherence to medical treatment: a review of reviews. *BMC Health Services Research, 7*, 55. https://doi.org/10.1186/1472-6963-7-55.

Veletsianos, G., & Doering, A. (2010). Long-term student experiences in a hybrid, open-ended and problem-based adventure learning program. *Australasian Journal of Educational Technology, 26*(2). https://doi.org/10.14742/ajet.1096.

Wampler, M. A., Galantino, M. L., Huang, S., Gilchrist, L. S., Marchese, V. G., Morris, G. S., et al. (2012). Physical activity among adult survivors of childhood lower-extremity sarcoma. *Journal of Cancer Survivorship: Research and Practice, 6*(1), 45–53. https://doi.org/10.1007/s11764-011-0187-5.

Wright, M. J., Galea, V., & Barr, R. D. (2005). Proficiency of balance in children and youth who have had acute lymphoblastic leukemia. *Physical Therapy, 85*, 782–790.

CHAPTER 5

Mindfulness techniques in the education of oral health professionals for the prevention and better management of stress

Cecilia Nunes[a],*, Diana Pinheiro[a,b],*, Manuela Soares Rodrigues[a,c],*, and Patrícia Rodrigues[a],*

[a]SPLS—Portuguese Society of Health Literacy, Lisbon, Portugal
[b]WFCMS—World Federation of Chinese Medicine Societies, Beijing, China
[c]Egas Moniz – Higher Education Cooperative, CRL, Almada, Portugal

1 Introduction

1.1 What is *mindfulness* and what are its origins?

Mindfulness originated from ancient Eastern and Buddhist philosophy (Germer, Siegel, & Fulton, 2005) and is an English translation of the word "Sati" in Pali, which means "full attention" (Germer, 2004). This definition has been expanded to allow a greater understanding and better accessibility, and applicability of this practice (Pinheiro & Rodrigues, 2021) (Fig. 1).

A practical definition of mindfulness is the awareness that emerges by paying attention, on purpose, in the present moment, and non-judgmentally to the unfolding of experience moment by moment (Kabat-Zinn, 2003) (Fig. 2).

The practice of mindfulness allows the transversal benefit of the professional-patient relationship (Fig. 3).

1.2 Health benefits of mindfulness

Mindfulness contributes to achieving greater levels of individual, personal, and institutional resilience for oral health professionals (OHPs) (Krogh, Medeiros, Bitran, & Langer, 2019). Furthermore, there is a positive relationship between empathy, work engagement, and less emotional exhaustion (Conversano et al., 2020).

Mindfulness increases the individual's ability to judge the situation more objectively and of a positive value. It is based on the ability to control and orientate attention, with an improvement in efficiency (Conversano et al., 2020).

---

* The authors have equally contributed to this article.

Active Learning for Digital Transformation in Healthcare Education, Training and Research
https://doi.org/10.1016/B978-0-443-15248-1.00004-7

65

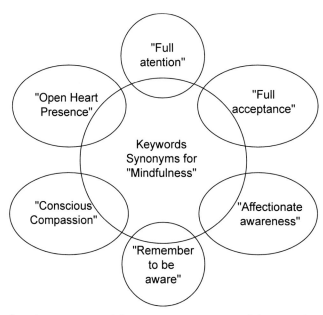

Fig. 1 Keywords found in ancient Buddhist texts. *(Source: Own elaboration, based on Germer, C., Siegel, R., & Fulton, P. (Eds.) (2005).* Mindfulness and psychotherapy. *New York: Guilford Press.)*

Fig. 2 Elements present at the same time in mindfulness. *(Source: Self-elaboration, based on Germer, C., Siegel, R., & Fulton, P. (Eds.) (2005).* Mindfulness and psychotherapy. *New York: Guilford Press.)*

It is necessary to have ongoing mindfulness practice programs for health professionals and to integrate this concept into the curriculum of health professionals (Conversano et al., 2020; Krogh et al., 2019).

Mindful practices have also been shown to provide benefits on aging in the Bio-Psycho-Social dimensions (Pinheiro & Rodrigues, 2021).

Improved professional care improves patient safety due to a more conscious practice and, consequently, a more positive therapeutic relationship (Vaz de Almeida, Rodrigues, Rodrigues, Pinheiro, & Nunes, 2021). Rodrigues (2021) reported that mindfulness is an activity that promotes the well-being of OHP to deal with occupational stress (Chart 1).

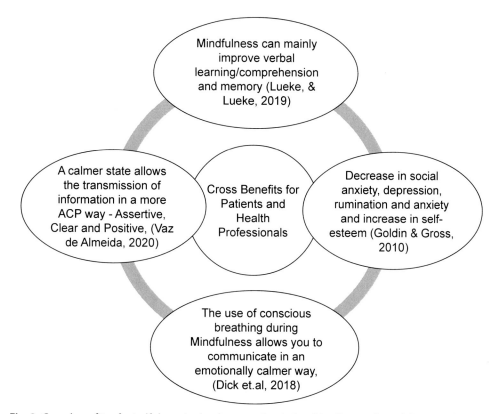

Fig. 3 Cross benefits of mindfulness in the therapeutic relationship. *(Source: Own elaboration based on the referred authors.)*

Chart 1 Results of the practice of mindfulness in health professionals.

Positive effects

Reduced anxiety, depression, and stress experiences	Less psychological symptoms, less anxiety, and less depression
Increased levels of mindfulness and self-compassion	Higher levels of individual, personal, and institutional resilience
Direct and positive impact on the well-being of doctors	Reduced likelihood of anxiety and burnout
Increased self-care	Less judgmental attitudes
Increased awareness	Less reactivity to difficulty
Concern focused on the other	Provides balance to the professional
Focuses the professional on the patient and guides them to action	Provides techniques for both the professional and the patient

Source: Rodrigues, P. (2021). Humor, well-being, and health literacy. In Vaz de Almeida, C., & Ramos, S. (Eds.), *Handbook of research on assertiveness, clarity, and positivity in health literacy*. IGI Global. https://doi.org/10.4018/978-1-7998-8824-6.

2 Stress level assessment instrument for oral health professionals

2.1 Application

This stress level assessment serves as an instrument for OHPs to consult and use to measure their perceived stress level in the clinical work environment.

Perceived stress is the feelings or thoughts that an individual has about how much stress they are under at a given point in time or over a given time period (Phillips, 2013).

A person's encounter with a stressful environment leads to a variety of coping processes, emotional states, disease precursors, and stress disorders (Lazarus, 1974). Every instance of adaptive commerce between a person and the environment is appraised cognitively as to its significance for the person's well-being. The self-regulatory processes as well as cognitive appraisals are key mediators of the person's reactions to stressful transactions and hence shape the somatic outcome (Lazarus, 1974).

2.2 Intervention

The assessment instrument is the Perceived Stress Scale (PSS-10), which focuses on simplicity, easy access, higher speed, and flexibility (Mujić Jahić, Bukejlović, Alić-Drina, & Nakaš, 2019).

The PSS is used with permission of the American Sociological Association, from Cohen, Kamarck, and Mermelstein (1983).

The PSS is a classic stress assessment instrument. The tool, though originally developed in 1983, remains a popular choice for helping us understand how different situations affect our feelings and our perceived stress. The questions in this scale are about professional feelings and thoughts during the last month. In each case, the professional is asked to indicate how often they felt or thought a certain way.

Perceived Stress Scale for Oral Health Professionals

Below are some questions about your professional feelings and thoughts during last month. Please choose from the following options:

0 = Never; 1 = Almost Never; 2 = Sometimes; 3 = Fairly Often; 4 = Very Often

1. In the last month, how often have you been upset because of something that happened unexpectedly?..............
2. In the last month, how often have you felt that you were unable to control the important things in your life? ....
3. In the last month, how often have you felt nervous and "stressed"? ....
4. In the last month, how often have you felt confident about your ability to handle your personal problems? ....
5. In the last month, how often have you felt that things were going your way?............
6. In the last month, how often have you found that you could not cope with all the things that you had to do? ....

7. In the last month, how often have you been able to control irritations in your life?...........

8. In the last month, how often have you felt that you were on top of things?............

9. In the last month, how often have you been angered because of things that were outside of your control?.............

10. In the last month, how often have you felt difficulties were piling up so high that you could not overcome them? ....

Figuring Your PSS Score

You can determine your PSS score by following these directions:

- First, reverse your scores for questions 4, 5, 7, and 8. For these four questions, change the scores in the following manner:
  $0=4, 1=3, 2=2, 3=1, 4=0$.
- Now add up your scores for each item to get a total. My total score is _____.
- Individual scores on the PSS can range from 0 to 40, with higher scores indicating higher perceived stress.
- Scores ranging from 0 to 13 would be considered low stress.
- Scores ranging from 14 to 26 would be considered moderate stress.
- Scores ranging from 27 to 40 would be considered high perceived stress.

Disclaimer: The scores on the following self-assessment do not reflect any particular diagnosis or course of treatment. They are meant as a tool to help assess your level of stress. If you have any further concerns about your current well-being, we advise you to contact a specialist.

Although the PSS-10 is a popular measure, a review of the literature by Taylor (2015) reveals three significant gaps: (a) There is some debate as to whether a one- or two-factor model best describes the relationships among the PSS-10 items; (b) little information is available on the performance of the items on the scale; and (c) it is unclear whether PSS-10 scores are subject to gender bias.

These gaps were addressed in Taylor's (2015) study and measurement invariance tests suggest that PSS-10 scores may not be substantially affected by gender bias. Overall, the findings suggest that inferences made using PSS-10 scores are valid.

3 Non-pharmacological strategies/tools for stress management

There are some non-pharmacological tools to help OHP to manage their stress. The following strategies can be carried out alone, combined, or sequentially integrated into stress reduction programs.

3.1 Breathing techniques

Mindfulness breathing, or mindfulness of breathing, is the physiological process transversal to all other mindful techniques and systems, which, when carried out in a slower and

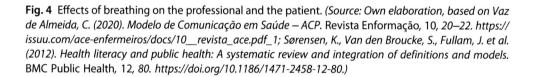

Healthcare professional

Listen and diagnose more effectively

Communicating in an emotionally calmer way, (Dick, T. E et.at 2018) is accessible and understandable.

Transmitting health information in a more ACP way - Assertive, Clear and Positive, (Vaz de Almeida, C. 2020)

Patient

Benefits from the confidence-boosting effect: "I am being listened to carefully and by a professional who appears to be calm. I can trust this health professional."

Access and understand more immediately, the health information received.

By better understanding the information, you will be able to make "good and correct health decisions" (Sorensen et al. 2012)

Fig. 4 Effects of breathing on the professional and the patient. *(Source: Own elaboration, based on Vaz de Almeida, C. (2020). Modelo de Comunicação em Saúde – ACP. Revista Enformação, 10, 20–22. https:// issuu.com/ace-enfermeiros/docs/10__revista_ace.pdf_1; Sørensen, K., Van den Broucke, S., Fullam, J. et al. (2012). Health literacy and public health: A systematic review and integration of definitions and models. BMC Public Health, 12, 80. https://doi.org/10.1186/1471-2458-12-80.)*

more profound way, allows for health benefits to be obtained at the respiratory, cardio-vascular, cardiorespiratory, and nervous levels (Russo, Santarelli, & O'Rourke, 2017) (Fig. 4).

The process can be observed before, during, and after the consultation (Fig. 5).

3.2 Muscle relaxation techniques

3.2.1 Progressive muscle relaxation (PMR)

Progressive muscle relaxation (PMR) is a fundamental form of stress management that was developed in the early 1920s by Edmund Jackson (Bracke, 2010). The procedure has been widely, and successfully, used to manage and treat a variety of anxiety disorders (Conrad & Roth, 2007).

The process of PMR requires the patient to focus on specific voluntary muscles and, in sequence, tense and then relax the tension in that muscle. As the tensing and relaxing sequence progresses, other aspects of the relaxation response also naturally occur: breathing becomes slower and deeper, heart rate and blood pressure decline, and vasodilatation in the small capillaries of the extremities may occur, creating a subjective sense of calmness and ease (Bracke, 2010).

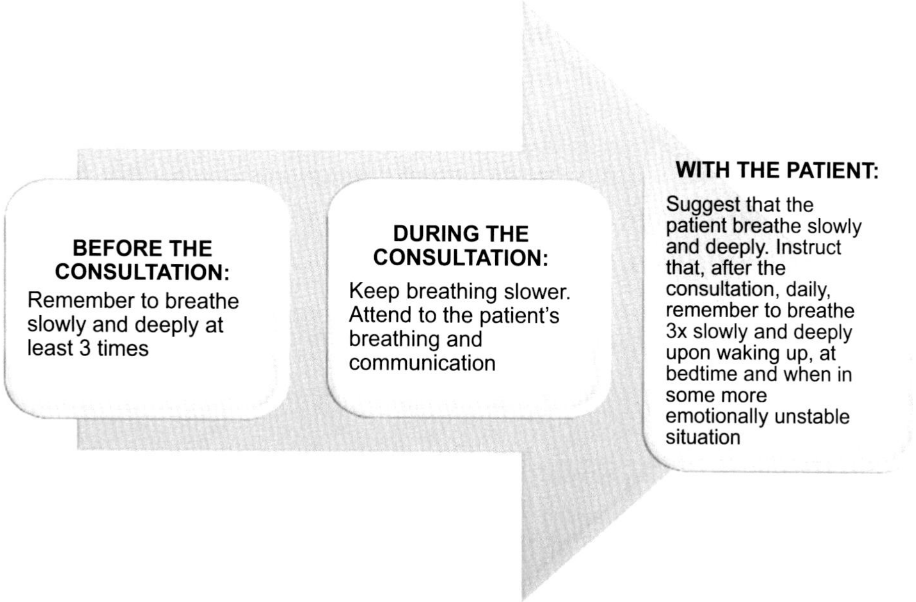

BEFORE THE CONSULTATION:
Remember to breathe slowly and deeply at least 3 times

DURING THE CONSULTATION:
Keep breathing slower. Attend to the patient's breathing and communication

WITH THE PATIENT:
Suggest that the patient breathe slowly and deeply. Instruct that, after the consultation, daily, remember to breathe 3x slowly and deeply upon waking up, at bedtime and when in some more emotionally unstable situation

Fig. 5 Breathing in clinical practice. *(Source: Own preparation inspired by Nunes (2007).* Comunicação em contexto clínico. *Lisboa: Bayer Health Care.)*

3.2.2 Body scan

The body scan is a somatically oriented, attention-focusing practice first introduced into clinical practice as part of the MBSR program developed by Jon Kabat-Zinn (Dreeben, Mamberg, & Salmon, 2013). In practice, participants begin the body scan by sitting or lying in a comfortable position. The instructor (live in class, then on audio recording at home) slowly guides the participants' attention through the various regions of the body (Dreeben et al., 2013). Compared to Western somatic practices such as PMR, the body scan is markedly less action-based and less goal-oriented and remains unique due to its focus on non-striving awareness without doing and its roots in Buddhist psychology (Dreeben et al., 2013).

3.3 Meditation

Mediation can be classified as formal meditation and informal meditation (Germer, 2004). The formal practice involves taking time to go to the mental "gym" or "temple". Regularly, set a schedule and dedicate a certain period to adopt the posture of sitting quietly, in meditation, or in the case of walking or moving meditation, repeat a sequence of movements (Chart 2).

Informal meditation, also called "Daily Mindfulness, can be practiced throughout the day, and it can be practiced before work life.

Chart 2 Practical examples of formal meditation.

1. Sitting meditation	Sitting relaxed and alert for several minutes, just "being" and experiencing the stillness. Practice being aware of thoughts, emotions, and sensations with a curious, open, compassionate, and non-judgmental mind
2. Meditation focusing on breathing	It consists of focusing attention on the breath and on the air entering through the nose and exiting through the lips. Breathing is also used in other types of meditation to get back to it each time the mind wanders.
3. Guided Meditation	It can be practiced sitting or lying down. It consists of the use of music or narrations that have the function of explaining how to manage breathing and relax.
4. Walking meditation	Paying attention to the feet when in contact with the surface of the earth with each step and noticing the sensations in the body. For each step you walk, you can look at your feet and inhale and exhale multiple times.
5. Moving meditation or mind-body exercises	It consists of sequences of exercises, normally relished slowly, which, transversally and from the beginning to the end of the practice, focus conscious attention on and require that you pay attention to the breath, physical sensations, and mental-emotional states. They may include stretching, continuous joint mobilization movements, or more static postures. Examples are Yoga, Therapeutic Qigong, and Tai Chi

Source: Own elaboration adapted from Germer, C. (2004). What is mindfulness? *Insight Journal, 22*(3), 24–29. Retrieved from: http://www.mindfulselfcompassion.org/articles/insight_germermindfulness.pdf; Hanh, T. N. (2014). *Meditação andando: Guia para a Paz Interior*, 21st ed. Editora Vozes; Kabat-Zinn, J. (2013). *Full catastrophe living: How to cope with stress, pain and illness using mindfulness meditation*. New York: Bantam Books. Paperback Edition. Song, R., Grabowska, W., Park, M., Osypiuk, K., Vergara-Diaz, G. P., Bonato, P., Hausdorff, J. M., Fox, M., Sudarsky, L. R., Macklin, E., & Wayne, P. M. (2017). The impact of tai chi and Qigong mind-body exercises on motor and non-motor function and quality of life in Parkinson's disease: A systematic review and meta-analysis. *Parkinsonism & Related Disorders, 41*, 3–13. https://doi.org/10.1016/j.parkreldis.2017.05.019).

It's about remembering throughout the day to pay attention to what's happening at the moment, without radically altering your routines (Chart 3).

In addition to the known benefits of mindful meditation on the burnout of health professionals, when practiced during periods of more prolonged stress, as was seen during the COVID-19 pandemic, it can contribute to a reduction in anxiety, depression, and pain (Behan, 2020).

3.4 Mind-body exercises

3.4.1 Yoga

Yoga involves physical poses, concentration, and deep breathing. Regular yoga practice can promote endurance, strength, calmness, flexibility, and well-being (U.S. Department of Health, and Human Services, National Center for Complementary Integrative Heath, 2020).

Chart 3 Practical examples of informal meditation or daily mindfulness.

1. Chewing	Chew slower and identify food flavors and textures.
2. Mobile phone	When the cell phone rings, try to listen first, paying attention to the tone and rhythm of the sound, as if you were listening to a musical instrument instead of being alert.
3. Sensory perception and temperature	Identify pleasant and unpleasant sensations and temperature in the feet while walking or sitting.
4. Textures	When picking up objects, feel their texture and present aromas, without traveling to memories of the past associated with that object, maintaining mindfulness in the moment.

Source: Own elaboration adapted from Germer, C. (2004). What is mindfulness? *Insight Journal, 22*(3), 24–29. Retrieved from: http://www.mindfulselfcompassion.org/articles/insight_germermindfulness.pdf; Kabat-Zinn, J. (2013). *Full catastrophe living: How to cope with stress, pain and illness using mindfulness meditation.* New York: Bantam Books. Paperback Edition; Song, R., Grabowska, W., Park, M., Osypiuk, K., Vergara-Diaz, G. P., Bonato, P., Hausdorff, J. M., Fox, M., Sudarsky, L. R., Macklin, E., & Wayne, P. M. (2017). The impact of tai chi and Qigong mind-body exercises on motor and non-motor function and quality of life in Parkinson's disease: A systematic review and meta-analysis. *Parkinsonism & Related Disorders, 41*, 3–13. https://doi.org/10.1016/j.parkreldis.2017.05.019).

The mindfulness-based yoga intervention had a statistically significant effect on the health and well-being of nurses and other healthcare professionals, most specifically for measures of stress: perceived stress, burnout, vitality, sleep quality, serenity, and mindfulness (Zhang, Song, Jiang, Ding, & Shi, 2020).

3.4.2 Qigong

Qigong is an ancient Chinese mindful practice that involves meditation, controlled breathing, and movement exercises. These involve simple, slow movements done repeatedly, combined with mindful breathing and self-massage.

Zou et al. (2017) show that Baduanjin Qigong practice is beneficial for quality of life, sleep quality, balance, handgrip strength, trunk flexibility, systolic and diastolic blood pressure, and resting heart rate. Wu et al. (2019) concluded that Qigong can reduce stress and anxiety after 2 to 6 months of practice. The findings from Guan, Hao, Guan, and Wang (2020) show that, compared with control interventions, *Baduanjin, a system* of Qigong exercises, seems to be an effective physical exercise in treating essential hypertension. Griffith et al. (2008) suggest that short-term exposure to Qigong was effective in reducing stress in hospital staff too.

3.4.3 Tai chi

Tai Chi is a gentle, safe, and uncomplicated comprehensive exercise and can help regulate body homeostasis and build up a good physique (Pan et al., 2021). The essential principles include the mind integrated with the body; control of movements and breathing; generating internal energy, mindfulness, song (loosening 松), and jing (serenity 静). The ultimate purpose of tai chi is to cultivate the vitality or life energy within us to flow smoothly and powerfully throughout the body.

Many of the upper limb movements are like those practiced in the Qigong system. In Qigong, there is essentially more expressive movement at the level of the upper limbs; in Tai Chi, the movement of the legs is added.

Though the researchers combined Qigong and Tai Chi studies in their review, they did find various positive results suggesting that both forms of exercise improve bone health and balance, as well as in terms of mental health, well-being, and quality of life. The results obtained by Wu et al. (2021) and Pan et al. (2021) indicate that the regular practice of Tai Chi is a conscious practice and is considered an antihypertensive lifestyle therapy.

3.5 Self-care

Most healthcare professionals are trained to put patients first. Self-care is not always prioritized among clinicians, as they may fear judgment from others or feel selfish at the thought of attending to their own needs. Practicing self-care could, however, be imperative to cope with the obligations, workload, and demands of their profession and help protect their health, well-being, and satisfaction with both their work and overall life (Søvold et al., 2021).

Recent reviews (Heath, Sommerfield, & von Ungern-Sternberg, 2020; Waris Nawaz, Imtiaz, & Kausar, 2020) recommend self-care as the first line of defense for healthcare workers to manage COVID-19 patient care demands. In this regard, these reviews highlight the importance of using supportive tools and techniques to combat mental health issues and compassion fatigue among healthcare workers, such as spiritual practices and relaxation techniques.

De Hert (2020) suggests that if the symptoms of burnout are minor and slight, measures such as changing life habits and optimizing work-life balance are recommended. These measures concentrate on three important pillars: relief from stressors, recuperation via relaxation and sport, and "return to reality" in terms of abandoning the ideas of perfection.

3.5.1 Enhancement of mood and positive emotions

The reinforcement of mood and positive emotions reduces the domain of negative emotions in the mind and body and improves the professional-patient relationship by increasing individual, personal, and institutional resilience (Vaz de Almeida et al., 2021). Humor is an essential part of our lives and an important way of dealing with stressful life events (Stieger, Formann, & Burger, 2011).

In a sample of pediatric dentists, a study was carried out on the use of humor in the clinical environment where dentists use common and individual types of humor and thus create a playful/humorous atmosphere for their patients (Nevo & Shapira, 1986).

Chart 4 MBSR program.

Session No.	Themes and activities
1	Introduction to Mindfulness
2	Perception
3	Mindfulness of breathing and the moving body
4	Learning about stress reaction patterns
5	Learning to deal with stress: use mindfulness to respond instead of react
6	Stressful communications and interpersonal mindfulness
7	How to take better care of me
8	Maintain mindfulness

The program consists of 8 weeks, with a weekly session of 2.5 h and an intensive day. It is fundamentally practical with group exercises and guided mindfulness meditation, including body scan, mindful breathing, and seated meditations. Source: Own elaboration adapted from Full Catastrophe Living, how to cope with stress, pain and illness using mindfulness meditation, 2013 Jon Kabat-Zinn, in Pinheiro, D., & Rodrigues, M. (2021). Os Benefícios do Mindfulness para um envelhecimento saudável e sustentável. *JIM - Jornal De Investigação Médica*, 2(2), 036–052. https://doi.org/10.29073/jim.v2i2.426.

The ability to identify the specific emotion that underlies the perception of stress is important, as the strategies necessary to deal with the stress caused by frustration with a difficult patient and that caused by the guilt and anxiety associated with having made a clinical error are likely to be different.

3.6 Mindfulness-based stress reduction programs

An MBSR is a synergistic articulation of mindfulness meditation, body awareness, and conscious movement (Raj & Kumar, 2018) (Chart 4).

According to Goodman and Schorling (2012), "A continuing education course based on mindfulness–based stress reduction was associated with significant improvements in burnout scores and mental well-being for a wide range of health professionals."

4 Stress triggers

4.1 Practice settings

OHPs report high levels of stress and increased burnout. Multiple forms of stressors lead to such occupational stress (Mujić Jahić et al., 2019). Lack of career perspective was the most crucial aspect in the development of burnout (Rada & Johnson-Leong, 2004).

Dealing with patients' emotions and discomfort is one of the major stress causes for OHPs (Myers & Myers, 2004; Rodrigues, 2021), increasing the time and effort needed to complete the treatment (D'Arro, 2018).

Consultation time, technical and personnel problems, and the number of hours worked per week are some of the factors of this stress (Myers & Myers, 2004). All these factors tend to affect the physical and mental health of OHPs, as well as have effects on personal relationships, professional relationships, health, and well-being (Rada & Johnson-Leong, 2004).

Certain stressors can stimulate people to grow professionally and personally, learn, or improve. The same stressors that are challenging in a positive sense also may be debilitating if they accumulate too rapidly (Rada & Johnson-Leong, 2004).

The operatory usually is small, and the dentist's focus is on an even smaller space, the oral cavity. Dentists are required to sit still for much of their workday, making very precise and slow movements with their hands while their eyes remain focused on a specific spot (Rada & Johnson-Leong, 2004).

A study of 306 dentists found that 90% of dentists feel body tension and only 54% use deliberate techniques to calm themselves (Rodrigues, 2021). Moreover, 50% said they have difficulties sleeping (Vaz de Almeida et al., 2021). These symptoms often are associated with stress.

4.2 Use (abuse) of technology

Technological evolution has brought remarkable contributions to the development of man in his social, cultural, and biological context; however, it has also been accompanied by numerous problems, including at work (Camelo & Angerami, 2008).

The introduction and implementation of new technologies in organizations have contributed to the emergence of situations that induce occupational stress due to the worker's need to adapt to the work process that requires new knowledge and skills (Martins, 2012).

The *technostress* model has gained immense popularity (Bondanini, Giorgi, Ariza-Montes, Vega-Munoz, & Andreucci-Annunziata, 2020), and all techno-stressors combined were associated with increased burnout (Kim, Lee, Yun, & Im, 2015).

4.3 Decision making

Stress and anxiety can both influence risk-taking in decision-making, with implications for patient safety (Croskerry, Abbass, & Wu, 2010).

While stress typically increases risk-taking, anxiety often leads to risk-averse choices (Hengen & Alpers, 2021).

Decision-making can be improved despite the effects of stress by implementing cognitive aids (e.g., checklists) and by teaching clinicians to better manage the deleterious effects through stress management or meditation training (Groombridge, Kim, Maini, Smit, & Fitzgerald, 2019) (Fig. 6).

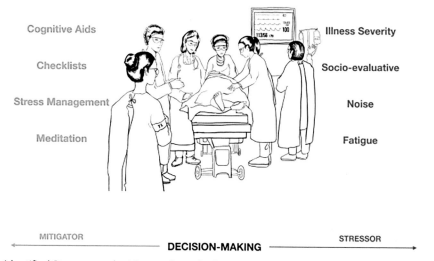

Cognitive Aids

Checklists

Stress Management

Meditation

Illness Severity

Socio-evaluative

Noise

Fatigue

MITIGATOR ← DECISION-MAKING → STRESSOR

Fig. 6 Identified Stressors and mitigators from the literature. *(Source: Groombridge, C., Kim, Y., Maini, A., Smit, D., & Fitzgerald, M. (2019). Stress and decision-making in resuscitation: A systematic review. Resuscitation, 144. https://doi.org/10.1016/j.resuscitation.2019.09.023.)*

5 Conclusion

Mindfulness techniques, practices, and systems can provide OHP with skills to manage stress and burnout, with benefits in their professional and personal activities.

Health benefits of mindfulness, when practiced during periods of more prolonged stress, are to be obtained at the respiratory, cardiovascular, cardiorespiratory, and nervous levels.

Total harmony of the inner and outer self comes from the integration of mind and body, which is one of the mottos of mindful practices.

Self-care strategies include spiritual practices and relaxation techniques and are important to deal with stressful situations. The reinforcement of mood and positive emotions improves the professional–patient relationship and leads to better clinical decision-making.

Despite the increasing use of mindfulness by OHPs, there is still some way to go. The support and reinforcement of these techniques by organizations is essential.

6 Added information (active learning)

1. **Suggested teaching assignments**
   (a) How many oral health professionals exist in your country? What percentage of them are using mindfulness techniques to reduce stress?

(b) Do public universities have "mindfulness" programs for students? If not, do you have any ideas to implement this?

(c) Mindfulness courses and breathing exercises as part of oral health professionals' continued education.

2. Recommended readings

Books:

- Alan Johnson, J. (2000). *Chinese medical Qigong therapy: A comprehensive clinical guide.* International Institute of Medical Qigong.
- Goleman & Davidson. (2018). *Livros Livros Técnicos Livros de Ciências Sociais e Humanas Livro.*
- Kabat-Zinn, J. (2013). *Full Catastrophe Living: How to Cope With Stress, Pain and Illness Using Mindfulness Meditation.* New York: Bantam Books. Paperback Edition.
- Stahl, B., & Goldstein, E. (2019). *A Mindfulness-Based Stress Reduction Workbook.* New Harbinger Publications.
- Swanson, A. (2019). *Understand the Anatomy and Physiology to Perfect Your Practice.* Dorling Kindersley.
- Wayne, P. (2013). *The Harvard Medical School Guide to Tai Chi: 12 Weeks to a Healthy Body, Strong Heart, and Sharp Mind.* Harvard Health Publications.

Websites with additional reading:

- https://www.helpguide.org/harvard/benefits-of-mindfulness.htm
- https://positivepsychology.com/mindfulness-based-stress-reduction-mbsr/
- https://learningresources.sjrstate.edu/Mindfulness/Meditation
- https://hr.harvard.edu/files/humanresources/files/mindfulness_now_and_zen.pdf

3. Case study

Description: In the pandemic context that we are now experiencing, and being that vaccination for COVID-19 is not mandatory, we start to see discriminatory and conflicting situations, leading to a huge stress level. This is applicable to public and shared spaces, schools, hospitals, cultural events, malls, sports, work, and many more.

(a) How can one reduce the level of stress when feeling discriminated against?

(b) Do any of the mindfulness techniques help people to feel more accepted by others?

(c) There are different types of stressed people - the ones that are almost permanently stressed and the ones that feel high stress associated with a specific situation/condition. COVID-19 is really a stress factor, either directly or indirectly. Would a COVID-19-specific mindfulness training be worth it?

4. Titles for research essays

(a) "Mindfulness techniques and oral health professionals stress—does it work?"

(b) "Patient—oral health professional relationship—techniques to gain confidence."

(c) "Can the mood/humor be considered an effective stress management strategy?"

5. **Recommended projects URL**
   (a) **Mindful Dentistry Training:** https://www.mindfuldentistrytraining.com/
   (b) **Soft Bites Podcast:** https://open.spotify.com/show/7hq7FJuP1eOW AbyLsNBSOn

References

Behan, C. (2020). The benefits of meditation and mindfulness practices during times of crisis such as COVID-19. *Irish Journal of Psychological Medicine, 37*(4) 256–258. https://doi.org/10.1017/ipm.2020.38.

Bondanini, G., Giorgi, G., Ariza-Montes, A., Vega-Munoz, A., & Andreucci-Annunziata, P. (2020). Technostress dark side of technology in the workplace: A scientometric analysis. *International Journal of Environmental Research and Public Health, 17*, 8013. https://doi.org/10.3390/ijerph17218013.

Bracke, P. E. (2010). Progressive muscle relaxation. In I. B. Weiner, & W. E. Craighead (Eds.), *In the Corsini encyclopedia of psychology*. https://doi.org/10.1002/9780470479216.corpsy0712.

Camelo, S. H. H., & Angerami, E. L. S. (2008). Riscos psicossociais no trabalho que podem levar ao estresse. *Revista Ciência cuidado e saúde, 7*, 232–240. Disponível em http://www.periodicos.uem.br/ojs/index.php/CiencCuidSaude/article/view/5010/3246.

Cohen, S., Kamarck, T., & Mermelstein, R. (1983). A global measure of perceived stress. *Journal of Health and Social Behavior, 24*, 386–396.

Conrad, A., & Roth, W. T. (2007). Muscle relaxation therapy for anxiety disorders: It works but how? *Journal of Anxiety Disorders, 21*(3) 243–264.

Conversano, C., Ciacchini, R., Orrù, G., Di Giuseppe, M., Gemignani, A., & Poli, A. (2020). Mindfulness, compassion, and self-compassion among health care professionals: What's new? A systematic review. *Frontiers in Psychology, 11*, 1683. https://doi.org/10.3389/fpsyg.2020.01683.

Croskerry, P., Abbass, A., & Wu, A. W. (2010). Emotional influences in patient safety. *Journal of Patient Safety, 6*, 199–205.

D'Arro, C. (2018). Mindful dentist. *International Journal of Whole Person Care, 5*(2) 29–37.

De Hert, S. (2020). Burnout in healthcare workers: Prevalence, impact and preventative strategies. *Local and Regional Anesthesia, 13*, 171–183.

Dreeben, S., Mamberg, M., & Salmon, P. (2013). The MBSR body scan in clinical practice. *Mindfulness, 4*, 394–401. https://doi.org/10.1007/s12671-013-0212-z.

Germer, C. (2004). What is mindfulness? *Insight Journal, 22*(3) 24–29. Retrieved from: http://www.mindfulselfcompassion.org/articles/insight_germermindfulness.pdf.

Germer, C., Siegel, R., & Fulton, P. (Eds.). (2005). *Mindfulness and psychotherapy*. New York: Guilford Press.

Goodman, M. J., & Schorling, J. B. (2012). A mindfulness course decreases burnout and improves well-being among healthcare providers. *International Journal of Psychiatry in Medicine, 43*(2) 119–128. https://doi.org/10.2190/PM.43.2.b.

Griffith, J. M., Hasley, J. P., Liu, H., Severn, D. G., Conner, L. H., & Adler, L. E. (2008). Qigong stress reduction in hospital staff. *Journal of Alternative and Complementary Medicine (New York, N.Y.), 14*(8) 939–945. https://doi.org/10.1089/acm.2007.0814.

Groombridge, C., Kim, Y., Maini, A., Smit, D., & Fitzgerald, M. (2019). Stress and decision-making in resuscitation: A systematic review. *Resuscitation, 144*. https://doi.org/10.1016/j.resuscitation.2019.09.023.

Guan, Y., Hao, Y., Guan, Y., & Wang, H. (2020). Effects of Baduanjin exercise on essential hypertension: A meta-analysis of randomized controlled trials. *Medicine, 99*(32) e21577. https://doi.org/10.1097/MD.0000000000021577.

Heath, C., Sommerfield, A., & von Ungern-Sternberg, B. S. (2020). Resilience strategies to manage psychological distress among healthcare workers during the COVID-19 pandemic: A narrative review. *Anaesthesia, 75*(10) 1364–1371.

Hengen, K. M., & Alpers, G. W. (2021). Stress makes the difference: Social stress and social anxiety in decision-making under uncertainty. *Frontiers in Psychology*, *22*(12) 578293. https://doi.org/10.3389/fpsyg.2021.578293.

Kabat-Zinn, J. (2003). Mindfulness-based interventions in context: Past, present, and future. *Clinical Psychology: Science and Practice*, *10*(2) 144–156. https://doi.org/10.1093/clipsy.bpg016.

Kim, H. J., Lee, C. C., Yun, H., & Im, K. S. (2015). An examination of work exhaustion in the mobile enterprise environment. *Technological Forecasting and Social Change*, *100*, 255–266. https://doi.org/10.1016/j.techfore.2015.07.009.

Krogh, E., Medeiros, S., Bitran, M., & Langer, Á. I. (2019). Mindfulness y la relación clínica: pasos hacia una resiliencia en medicina. *Revista Medica de Chile*, *147*(5) 618–627.

Lazarus, R. S. (1974). Psychological stress and coping in adaptation and illness. *International Journal of Psychiatry in Medicine*, *5*(4) 321–333.

Martins, M. C. A. (2012). Situações indutoras de estresse no trabalho do enfermeiro em ambiente hospitalar. *Millenium - Revista do ISPV.*, *28*, 10. Disponível em: http://www.ipv.pt/millenium/millenium28/18.htm.

Mujić Jahić, I., Bukejlović, J., Alić-Drina, S., & Nakaš, E. (2019). Assessment of stress among doctors of dental medicine. *Acta Stomatologica Croatica*, *53*(4) 354–362.

Myers, H. L., & Myers, L. B. (2004). It's difficult being a dentist': Stress and health in the general dental practitioner. *British Dental Journal*, *197*(2) 89–101. https://doi.org/10.1038/sj.bdj.4811476.

Nevo, O., & Shapira, J. (1986). Use of humor in managing clinical anxiety. *ASDC Journal of Dentistry for Children*, *53*(2) 97–100.

Pan, X., Tian, L., Yang, F., Sun, J., Li, X., An, N., et al. (2021). Tai chi as a therapy of traditional Chinese medicine on reducing blood pressure: A systematic review of randomized controlled trials. *Evidence-based Complementary and Alternative Medicine: Ecam*, *2021*, 4094325. https://doi.org/10.1155/2021/4094325.

Phillips, A. C. (2013). Perceived stress. In M. D. Gellman, & J. R. Turner (Eds.), *Encyclopedia of behavioral medicine*. New York, NY: Springer. https://doi.org/10.1007/978-1-4419-1005-9_479.

Pinheiro, D., & Rodrigues, M. (2021). Os Benefícios do Mindfulness para um envelhecimento saudável e sustentável. *JIM - Jornal De Investigação Médica*, *2*(2) 036–052. https://doi.org/10.29073/jim.v2i2.426.

Rada, R. E., & Johnson-Leong, C. (2004). Stress, burnout, anxiety and depression among dentists. *Journal of the American Dental Association (1939)*, *135*(6) 788–794. https://doi.org/10.14219/jada.archive.2004.0279.

Raj, A., & Kumar, P. (2018). Efficacy of mindfulness based stress reduction (MBSR): A brief overview. *Journal of Disability Management and Rehabilitation*, *4*(1). https://www.researchgate.net/publication/328540330_Efficacy_of_Mindfulness_Based_Stress_Reduction_MBSR_A_Brief_Overiew.

Rodrigues, P. (2021). Humor, well-being, and health literacy. In C. Vaz de Almeida, & S. Ramos (Eds.), *Handbook of research on assertiveness, clarity, and positivity in health literacy* IGI Global. https://doi.org/10.4018/978-1-7998-8824-6.

Russo, M. A., Santarelli, D. M., & O'Rourke, D. (2017). The physiological effects of slow breathing in the healthy human. *Breathe (Sheffield, England)*, *13*(4) 298–309. https://doi.org/10.1183/20734735.009817.

Søvold, L. E., Naslund, J. A., Kousoulis, A. A., Saxena, S., Qoronfleh, M. W., Grobler, C., et al. (2021). Prioritizing the mental health and well-being of healthcare workers: An urgent global public health priority. *Frontiers in Public Health*, *9*, 679397.

Stieger, S., Formann, A. K., & Burger, C. (2011). Humor styles and their relationship to explicit and implicit self-esteem. *Personality and Individual Differences*, *50*(5) 747–750. https://doi.org/10.1016/J.PAID.2010.11.025.

Taylor, J. M. (2015). Psychometric analysis of the ten-item perceived stress scale. *Psychological Assessment*, *27*(1) 90–101.

U.S. Department of Health & Human Services, National Center for Complementary Integrative Heath. (2020). *Yoga for Health: What the Science Says*. https://www.nccih.nih.gov/health/providers/digest/yoga-for-health-science.

Vaz de Almeida, C., Rodrigues, P., Rodrigues, M., Pinheiro, D., & Nunes, C. (2021). *Bem-Estar, Mindfulness e Saúde Oral: Benefícios para Profissionais e Pacientes*. Lisboa: APPSP. https://doi.org/10.5281/zenodo.4557438.

Waris Nawaz, M., Imtiaz, S., & Kausar, E. (2020). Self-care of frontline health care workers: During COVID-19 pandemic. *Psychiatria Danubina*, *32*(3–4) 557–562.

Wu, Y., Johnson, B. T., Chen, S., Chen, Y., Livingston, J., & Pescatello, L. S. (2021). Tai Ji Quan as anti-hypertensive lifestyle therapy: A systematic review and meta-analysis. *Journal of Sport and Health Science*, *10*(2) 211–221. https://doi.org/10.1016/j.jshs.2020.03.007.

Wu, J. J., Zhang, Y. X., Du, W. S., Jiang, L. D., Jin, R. F., Yu, H. Y., et al. (2019). Effect of qigong on self-rating depression and anxiety scale scores of COPD patients: A meta-analysis. *Medicine*, *98*(22) e15776. https://doi.org/10.1097/MD.0000000000015776.

Zhang, X. J., Song, Y., Jiang, T., Ding, N., & Shi, T. Y. (2020). Interventions to reduce burnout of physicians and nurses: An overview of systematic reviews and meta-analyses. *Medicine*, *99*(26) e20992. https://doi.org/10.1097/MD.0000000000020992.

Zou, L., SasaKi, J. E., Wang, H., Xiao, Z., Fang, Q., & Zhang, M. (2017). A systematic review and Meta-analysis Baduanjin qigong for health benefits: Randomized controlled trials. *Evidence-Based Complementary and Alternative Medicine: eCAM*, *2017*, 4548706. https://doi.org/10.1155/2017/4548706.

Further reading

Cohen, S., & Williamson, G. (1988). Perceived stress in a probability sample of the United States. In S. Spacapan, & S. Oskamp (Eds.), *The social psychology of health*. Newbury Park, CA: Sage.

Dick, T. E., Dutschmann, M., Feldman, J. L., Fong, A. Y., Hülsmann, S., Morris, K. M., et al. (2018). Facts and challenges in respiratory neurobiology. *Respiratory Physiology & Neurobiology*, *258*, 104–107. https://doi.org/10.1016/j.resp.2015.01.014.

Ditto, B., Eclache, M., & Goldman, N. (2006). Short-term autonomic and cardiovascular effects of mindfulness body scan meditation. *Annals of Behavioral Medicine: A Publication of the Society of Behavioral Medicine*, *32*(3) 227–234.

Goldin, P. R., & Gross, J. J. (2010). Effects of mindfulness-based stress reduction (MBSR) on emotion regulation in social anxiety disorder. *Emotion (Washington, D.C.)*, *10*(1) 83–91. https://doi.org/10.1037/a0018441.

Lazarus, R. S., & Stress, F. S. (1984). *Appraisal, and coping*. New York: Springer Publishing Company.

Lueke, A., & Lueke, N. (2019). Mindfulness improves verbal learning and memory through enhanced encoding. *Memory & Cognition*, *47*(8) 1531–1545. https://doi.org/10.3758/s13421-019-00947-z.

CHAPTER 6

Caring for a woman's sleep through interpersonal teaching in proximity medicine consultations

Marta Barroca
Group of Health Centers (ACES) of the Tagus Estuary, Administration Regional Health (ARS) of Lisbon and Tagus Valley
Faculty of Medicine of Lisbon
Health Literacy from Ispa—University Institute

1 The importance of sleep

Sleep disorders are among the top 10 reasons for primary care consultations and among the top 20 of the most frequent diagnoses in family medicine, having a huge impact on health and on the economy. People with sleep disorders tend to have more consultations, consume more medications, take more complementary diagnostic tests, and be hospitalized more often. Facing this reality, there should be greater awareness to systematically question sleep habits in clinical consultations, as already done for nutrition habits or physical activity (Mateus, 2006; Rodrigues, Nina, & Matos, 2014). Likewise, robust campaigns must be promoted to raise awareness of the importance of sleep in health and to improve people's sleep literacy and sleep habits (Reis et al., 2018).

Sleep is a biological and behavioral phenomenon essential to human homeostasis at all stages of the life cycle (Frange et al., 2017). Humans spend about a third of their lives sleeping, which denotes the importance of sleep for daily functioning, health, and well-being (van de Straat & Bracke, 2015). Sleeping is not the opposite of being awake, because it is during sleep that the human body carries out numerous complex and indispensable functions, such as cell growth and regeneration, regulation of metabolism and the immune system, energy conservation, learning and memorization, physical and cognitive performance renewal, mood regulation, everyday "offline processing", and synapse regulation. These processes occur at the expense of abundant neuronal activity and the production of fundamental anabolic hormones (growth hormone, cortisol, testosterone, prolactin, etc.), promoting health, well-being, and quality of life (Reis et al., 2018).

Sleep is influenced by several internal and external factors, such as genetics, behaviors, comorbidities, physical environment, and society; it has individual variability and changes throughout the different stages of the life cycle and according to gender. However, to be considered good quality, sleep must have adequate duration and time, regularity, and absence of sleep disturbances or diseases. Optimal sleep duration has been

Active Learning for Digital Transformation in Healthcare Education, Training and Research
https://doi.org/10.1016/B978-0-443-15248-1.00003-5

widely discussed and varies by individual and even by geographical region. American recommendations point to 7 to 9 h of sleep per night as the ideal duration of sleep in adults (Watson et al., 2015). However, the demands of today's society, where there is no longer a clear demarcation between day and night—altering the circadian rhythm that regulates sleep—, where there is wide use of screens, and where shift work and professional/domestic tasks accumulate, lead to serious consequences on the quantity and quality of sleep for both men and women of all ages. In addition, aging itself leads to changes in sleep architecture, predisposing to various disorders (Frange et al., 2017; Logan & McClung, 2019). Other factors associated with these disorders are stress, alcohol or drug use, depression, low educational or socioeconomic status, marital status (divorced, separated, widowed), and female gender (van de Straat & Bracke, 2015). Even so, surrounded by all these social, professional, and cultural demands, most people devalue the importance of sleep and are not sufficiently aware of the health risks of sleep deprivation and sleep disturbances. Less sleep is related to worse health outcomes, which can have serious or even fatal consequences, namely an increase in the chances of obesity, diabetes, cardiovascular diseases, dementia, and accidents (Tao, Sun, Shao, Li, & Teng, 2016).

2 Women's sleep

Being a woman is a risk factor for sleep disorders and the impact of sleep disorders is different between men and women (Vézina-Im, Moreno, Thompson, Nicklas, & Baranowski, 2017). While the normal number of hours of daily sleep varies from 7 to 9 continuous hours, the average for women is 6 h and 40 min, with interruptions (National Sleep Foundation, 2011).

Women's sleep has different characteristics than male sleep, which is noticed as early as in puberty, due to the effect of female sex hormones in the brain, genetic mechanisms linked to the female sex, and specific female psychological and behavioral factors. Since the brain's sleep-regulating centers are sensitive to female sex hormones and these vary throughout a woman's life cycle, the quality and quantity of sleep also change through the different stages of life, namely menstrual cycle at childbearing age, pregnancy, puerperium, and peri- and post-menopause, predisposing to different sleep disorders (Bei, Coo, Baker, & Trinder, 2015; Pengo, Won, & Bourjeily, 2018).

Menarche marks the beginning of childbearing age, with an increase in the ovarian production of estrogen and progesterone—hormones that influence homeostatic functions at different levels: circulatory, respiratory, metabolic, and sleep-wake cycle. Progesterone promotes deep sleep and total sleep time, also acting as a positive respiratory stimulus and decreasing upper airway resistance, which protects against respiratory sleep disorders such as obstructive sleep apnea syndrome (OSAS). Estrogens also seem to have an important protective role against OSAS and its cardiovascular consequences, by decreasing cellular oxidative stress. On the other hand, female sex hormones are also related to psychological factors that influence mood and the woman's subjective perception of sleep quality, right from menarche. Although women report earlier and subjectively worse sleep quality than men, objectively, this has not been proven (Pengo et al., 2018).

Sleep disorders in women of childbearing age are common during the menstrual cycle: insomnia or premenstrual hypersomnolence, related to the luteal phase of the cycle, where there is a higher concentration of luteinizing hormone (LH) and progesterone. Parasomnias—involuntary and undesirable physical experiences (movements, behaviors, perceptions, dreams, emotions) that occur when falling asleep, during sleep, or when waking up—can also occur. OSAS, restless legs syndrome (RLS), and primary insomnia will be less common but also possible in this age group—sometimes with more subtle symptoms, leading to underdiagnosis and delay in treatment (Attarian & Viola-Saltzman, 2013).

Pregnant women undergo a series of dynamic physiological and anatomical changes throughout pregnancy, which affect important changes in sleep: change in duration, greater fragmentation, and respiratory and metabolic changes. There may be increased sleepiness in the first trimester, due to increased progesterone and basal temperature, although the onset of nocturia and muscle discomfort can negatively impact sleep already in this phase. In the second and third trimesters, it is common for sleep to be negatively affected by fetal movements, uterine contractions, rhinitis and nasal congestion, orthopnea, heartburn, muscle cramps, and body position (Pengo et al., 2018). In these phases, there is a greater predisposition to insomnia, OSAS, and RLS, which, despite generally regressing in the puerperium, increase the obstetric and fetal risk for gestational diabetes, hypertension in pregnancy, need for cesarean section, or preterm delivery (Attarian & Viola-Saltzman, 2013; Ölmez et al., 2015; Pengo et al., 2018). During labor, especially if it is nocturnal, and immediate postpartum, it is common to experience acute sleep deprivation, which is chronically prolonged in the first months of the baby's life, meeting the baby's needs of feeding and night care. Even though total maternal sleep time is not significantly affected in the first few months of the baby's life, the increase in sleep fragmentation is sufficient to decrease a woman's subjective sleep quality (Bei et al., 2015).

Sleep disorders are very prevalent in peri- and post-menopausal women (40%–60%). In this phase, there is greater sleep fragmentation, more nocturnal awakenings, less total sleep time, and less sleep efficiency, with a worse subjective perception of sleep quality, compared to younger women. The main causes for these changes are vasomotor symptoms and consequent night sweats, which are conditioned by hormonal changes (decreased estrogen and progesterone) and lead to insomnia. On the other hand, the decrease in estrogen also causes a redistribution of body fat, with enlargement of the neck and greater flaccidity of the cervical muscles, predisposing to OSAS. RLS, also very common at this age, can be idiopathic (no detectable cause) or be related to iron and dopamine deficiency—relatively common in older age or as a side effect of drugs. Aging and hormonal changes, including a decrease in melatonin production, are also responsible for the described changes in the sleep of older women and for disturbances in their circadian rhythm, such as advanced sleep phase syndrome. On the other hand, comorbidities such as depression and anxiety are very common at this age, which themselves lead to sleep alterations and are also influenced by sleep quality, in a bidirectional relationship (Attarian & Viola-Saltzman, 2013; de Campos, Bittencourt, Haidar, Tufik, & Baracat, 2005; Freeman, Sammel, Gross, & Pien, 2015; Pengo et al., 2018).

In conclusion, poor quality or quantity of sleep leads to excessive daytime sleepiness and fatigue, with a negative impact on women's overall health, mood, cognitive and work performance, social interactions, safety, and quality of life, also directly or indirectly affecting their children and family, given the intrinsically caring characteristic of women, at all stages of their life cycle.

3 Sleep interventions

When approaching women's sleep, health professionals must be aware of the multiplicity of factors involved (hormonal, physiological, psychological, familial, social); the heterogeneity of sleep disturbances that may arise throughout the life cycle; and the interdependence between sleep, physical, and mental health. Sleep disorders can often be the cause but also the consequence of other comorbidities, needing specific intervention (Bei et al., 2015).

Besides pharmacological interventions that will have their place and relevance in certain sleep disorders or associated comorbidities, it is extremely important to disseminate behavioral interventions that facilitate a good relationship between women and sleep and improve their literacy on the subject and sleep quality and health. This approach uses the premises of lifestyle medicine, in which changes in health behaviors and habits may have a therapeutic role in the prevention and treatment of non-communicable diseases, involving the individual in the process of change and care in an active, conscious, and participatory way.

A recent literature review distinguished and evaluated the evidence of 11 different types of sleep interventions in healthy people: sleep education (seminars, brochures, online); behavioral change techniques (positive routines, regular sleep schedule, scheduled awakenings, etc.); relaxation techniques (relaxation music, progressive relaxation, mindfulness); physical exercise (aerobic, Pilates, low-intensity exercise); mind-body exercise (physical activity combined with meditation – Tai Chi, yoga, Qigong); aromatherapy and massage; interventions in the physical environment (light, sound, temperature); psychotherapy; delay in starting school hours; combined interventions (e.g. sleep education and behavior change techniques); and other interventions (nutrition, infrared light, cryostimulation, acupuncture, hypnosis, biofeedback, etc.). The interventions that showed the most robust effectiveness evidence in improving sleep were later school hours, behavior change techniques, and mind–body exercise (Albakri, Drotos, & Meertens, 2021).

Psychological interventions that readjust women's negative cognitions about sleep help them to better understand the physiological changes they are experiencing and allow them to demystify beliefs, being very useful in improving their psychological well-being and consequently their sleep (Bei et al., 2015). Searching for positive experiences that give purpose to life; identifying automatic thoughts and replacing them with more positive and functional alternative responses; developing strategies to overcome obstacles; and doing realistic changes in daily routine/daily behaviors that lead to greater well-being, all these are strategies that improve levels of anxiety and depression and therefore may indirectly lead to better sleep quality (Friedman et al., 2019).

The physical environment affects sleep: noise, temperature, light, air quality, and mattress and pillow comfort. Brief, simple interventions that raise awareness of these topics can have positive results in sleep and insomnia (Desjardins, Lapierre, Vasiliadis, & Hudon, 2020).

4 Health literacy and communication

People with lower levels of health literacy are 1.5 to 3 times more likely to have poorer health outcomes. They will have more difficulty identifying health services, filling out health forms, sharing their medical history with the health professional, and relating behaviors and symptoms to diseases. It is therefore plausible that lower levels of health literacy lead to greater difficulty in describing symptoms related to sleep disorders, adhering to complementary sleep-related diagnostic tests, and adhering to proposed treatments—namely CPAP (continuous positive air pressure) for OSAS, which requires higher levels of literacy (Williams et al., 2016).

With better health literacy, individuals will be able to participate directly in preventive and health promotion measures and in early detection and better management of the disease, having a more informed and adequate access to health care. Better health literacy will improve adherence to treatments, with the ultimate aim of improving the quality of life. In this health literacy acquisition process, each individual is involved in the family and community, in a given biopsychosocial context, with a reciprocal interaction at each level of the system: intrapersonal (genetics, personality), peer (family, friends), organizational (health care, work, school), community, and socio-political (Bonuck, Blank, True-Felt, & Chervin, 2016; McLeroy, Bibeau, Steckler, & Glanz, 1988).

Health professionals have a guiding role and are fundamental agents for behavioral change and promotion of health literacy in the general population. For this purpose, they must master specific techniques of communication and health literacy tools.

Health communication is the "interpersonal or mass communication activity aimed at improving the health status of individuals and populations" (Nutbeam, 1998). The main errors that lead to poor health communication are lack of understanding of the contents; confusion in understanding the meaning of words; similar or very technical names of drugs or therapeutic products; misunderstanding of the guidelines given; unmonitored use of products; the health professional–patient relationship excessively technical and without interpersonal investment.

Due to the efficiency and productivity demands of the healthcare sector nowadays, health professionals have less and less time to communicate with their patients properly. In addition, medical education itself, historically paternalistic, promotes an asymmetrical relationship between professionals and patients. Agendas are different at each health meeting. All these factors contribute to communication errors and undermine literacy gains.

When approaching women's sleep in medical consultations, as approaching any other health issue, it is possible to use simple and effective techniques that create empathy, proximity, and a better doctor-patient relationship for the therapeutic process, with the perspective of improving individuals' literacy and ultimately their health and quality of life. Looking into the eyes, respecting each other, having a consonant and reflective body language, respecting pauses, giving space to silences, listening actively, asking open questions, having an assertive, clear, positive language (ACP model), creating physical proximity, and using therapeutic touch are key examples (Almeida, Moraes, & Brasil, 2020; Belim & de Almeida, 2018).

Another communication technique that emerged as a response to the limitations of the biomedical model, attempting to improve communication in the health context, is narrative medicine. Narrative medicine is based on storytelling—the patient tells his/her story, his/her perspective of illness/health; he/she is understood as the unique being that he/she is; he/she feels validated; there is mutual sharing of stories between health professional and the patient; and greater empathy and close communication are created. All these characteristics end up creating an intrinsically therapeutic effect in storytelling, with positive effects on the participants' well-being and on the motivation for change/action regarding the problem addressed. In the same way that this communication technique can be applied in a meeting between the health professional and the patient (in a consultation), it can also be used in peer groups and therapeutic groups. After all, it is through stories—storytelling—that the world acquires meaning for human beings and that they communicate what is truly relevant to them (Fioretti et al., 2016; Zaharias, 2018).

All these techniques are the core center of a mindful practice, where the doctor should be in the present moment, connected with the patient and with himself/herself, with a spirit of curiosity, without judgment, acting with compassion but technical competence, and making decisions based on scientific evidence, with respect to the values of both patient and doctor, in presence and with insight (Epstein, 1999).

5 Conclusion

Having in mind the theory about sleep's biology, functions, and disorders, the particularities of women's sleep all through the life cycle, and good communication techniques, a clinician takes good care of women's sleep, leading to the improvement in health literacy, sleep literacy, and quality of life.

6 Active learning sections

1. **Suggested teaching assignments**
   Try to deepen your knowledge about:
   A. the most frequent sleep disorders in women: insomnia, OSAS, RLS, etc.
   B. cognitive behavior therapy for insomnia
   C. lifestyle medicine concept
2. **Recommended readings**
   A. https://worldsleepday.org/usetoolkit/talking-points
   B. https://aasm.org/resources/pdf/adultsleepdurationconsensus.pdf
   C. https://www.aafp.org/family-physician/practice-and-career/managing-your-career/physician-well-being/practicing-self-care/mindfulness.html
   D. https://www.sleepfoundation.org/women-sleep/healthy-sleep-tips-women
3. **Case study**

A 51-year-old healthy female beautician lives with her partner and has a 25-year-old son who already lives on his own. She reported tiredness lasting more than 6 months, morning headaches that last all day, daytime hypersomnolence, important memory changes, and worse performance at work, without changes in appetite, weight, or mood. Some nights, she had initial insomnia. She described her daily routine and sleep schedules: she worked about 11 to 12 h a day, without time for herself and without physical exercise, and her excessive tiredness in the evening caused her to fall asleep on the couch for 1 to 2 h, which made it difficult to sleep at night - initial insomnia and fragmented sleep. She slept only 5 to 6 h a night. The physical examen was normal. Her blood tests showed a decrease in ferritin, with no other changes. The doctor explained to the patient that the symptoms presented might be due to chronic sleep deprivation and suggested better sleep hygiene and strategies to increase daily relaxation, as well as physical exercise and some time for leisure. Iron supplementation was started due to low ferritin and clonazepam 0.5 mg was prescribed in SOS, if needed to improve the initial insomnia. Two months later, on reassessment, the patient reported a significant reduction in fatigue and headaches and an improvement in cognitive functions. She managed to take daily walks and decided to work less than an hour a day. She tried to keep regular sleep schedules, as advised, and get at least 7 h of sleep per night. She no longer had periods of daytime hypersomnolence and she didn't use clonazepam, because initial insomnia reduced significantly.

Topics to reflect:

A. In your opinion, which communication techniques could have been used with this patient, in order to give the doctor a better understanding of the problem and an accurate diagnosis?

B. Can we say, from the results obtained, that the intervention of the health professional increased the sleep literacy of this patient?

C. What examples of practices based on lifestyle medicine were used in this case while promoting health literacy gains?

4. **Titles for research essays (a few indicative titles for possible research works related to the theme of the paper)**

A. Mindfulness for insomnia

B. Behavioral changes in infant's sleep and its effect on mother's sleep

C. Sleep in nursing homes—are there differences between institutionalized men and women? Strategies that work

5. **Recommended Projects URL**

A. American College of Lifestyle Medicine: https://lifestylemedicine.org/

B. Women and Sleep—An overview of various sleep disorders that on average affect women more: https://www.sleepfoundation.org/women-sleep

C. Create a Health Literacy Plan: https://www.cdc.gov/healthliteracy/planact/develop/index.html

D. Four Evidence-Based Communication Strategies to Enhance Patient Care: https://www.aafp.org/pubs/fpm/issues/2018/0900/p13.html

References

Albakri, U., Drotos, E., & Meertens, R. (2021). Sleep health promotion interventions and their effectiveness: An umbrella review. *International Journal of Environmental Research and Public Health*, *18*(11).

Almeida, C. V., Moraes, K., & Brasil, V. (2020). *50 Técnicas de literacia em saúde na prática. Um guia para a saúde*. Novas Edições Académicas.

Attarian, H. P., & Viola-Saltzman, M. (2013). *Sleep Disorders in Women*. Available from: http://link.springer.com/10.1007/978-1-62703-324-4.

Bei, B., Coo, S., Baker, F. C., & Trinder, J. (2015). Sleep in women: A review. *Australian Psychologist*, *50*(1), 14–24.

Belim, C., & de Almeida, C. V. (2018). Communication competences are the key! A model of communication for the health professional to optimize the health literacy – Assertiveness, clear language and positivity. *The Journal of Health Communication*, *03*(03), 1–13. Available from: http://healthcare-communications.imedpub.com/communication-competences-are-the-keya-model-of-communication-for-the-healthprofessional-to-optimize-the-health-literacy-assertive.php?aid=22761.

Bonuck, K. A., Blank, A., True-Felt, B., & Chervin, R. (2016). Promoting sleep health among families of young children in head start: Protocol for a social-ecological approach. *Preventing Chronic Disease*, *13*(7), 160144. Available from: http://www.cdc.gov/pcd/issues/2016/16_0144.htm.

de Campos, H. H., Bittencourt, L. R. A., Haidar, M. A., Tufik, S., & Baracat, E. C. (2005). Prevalência de distúrbios do sono na pós-menopausa. *Revista Brasileira de Ginecologia e Obstetrícia*, *27*(12), 731–736. Available from: http://www.scielo.br/scielo.php?script=sci_arttext&pid=S0100-72032005001200005&lng=pt&nrm=iso&tlng=pt.

Desjardins, S., Lapierre, S., Vasiliadis, H. M., & Hudon, C. (2020). Evaluation of the effects of an intervention intended to optimize the sleep environment among the elderly: An exploratory study. *Clinical Interventions in Aging*, *15*, 2117–2127.

Epstein, R. M. (1999). Mindful practice. *Journal of the American Medical Association*, *282*(9), 833–839.

Fioretti, C., Mazzocco, K., Riva, S., Oliveri, S., Masiero, M., & Pravettoni, G. (2016). Research studies on patients' illness experience using the narrative medicine approach: A systematic review. *BMJ Open*, *6*, e011220. https://doi.org/10.1136/bmjopen-2016-011220.

Frange, C., Banzoli, C. V., Colombo, A. E., Siegler, M., Coelho, G., Bezerra, A. G., et al. (2017). Women's sleep disorders: Integrative care. *Sleep Science*, *10*(4), 174–180.

Freeman, E. W., Sammel, M. D., Gross, S. A., & Pien, G. W. (2015). Poor sleep in relation to natural menopause: A population-based 14-year follow-up of midlife women. *Menopause*, *22*(7), 719–726.

Friedman, E. M., Ruini, C., Foy, C. R., Jaros, L., Love, G., & Ryff, C. D. (2019). Lighten UP! A community-based group intervention to promote eudaimonic well-being in older adults: A multi-site replication with 6 month follow-up. *Clinical Gerontologist*, *42*(4), 387–397.

Logan, R. W., & McClung, C. A. (2019). Rhythms of life: circadian disruption and brain disorders across the lifespan. In *Vol. 20. Nature reviews neuroscience* (pp. 49–65). Nature Publishing Group.

Mateus, A. (2006). *Em sonhos, diurnos pesadelos*.

McLeroy, K. R., Bibeau, D., Steckler, A., & Glanz, K. (1988). An_Ecological_Perspective_on_Health_Prom.pdf. *Health Education Quarterly*, *15*(4), 351–377.

National Sleep Foundation. (2011). *Women and sleep*. Available from: www.sleepfoundation.org.

Nutbeam, D. (1998). Health promotion glossary. *Health Promotion International*, *13*, 4.

Ölmez, S., Keten, H. S., Kardaş, S., Avcı, F., Dalgacı, A. F., Serin, S., et al. (2015). Factors affecting general sleep pattern and quality of sleep in pregnant women. *Journal of Turkish Society of Obstetric and Gynecology*, *12*(1), 1–5. Available from: http://cms.galenos.com.tr/Uploads/Article_10259/1-5.pdf.

Pengo, M. F., Won, C. H., & Bourjeily, G. (2018). Sleep in women across the life span. In *Vol. 154. Chest* (pp. 196–206). Elsevier Inc.

Reis, C., Dias, S., Rodrigues, A. M., Sousa, R. D., Gregório, M. J., Branco, J., et al. (2018). Sleep duration, lifestyles and chronic diseases: A cross-sectional population-based study. *Sleep Science*, *11*(4), 217–230.

Rodrigues, M., Nina, S., & Matos, L. (2014). Como dormimos? Avaliação da qualidade do sono em cuidados de saúde primários. *Revista Portuguesa de Medicina Geral e Familiar*, *30*(1), 16–22. Available from: http://www.scielo.mec.pt/scielo.php?pid=S2182-51732014000100004&script=sci_arttext&tlng=en.

Tao, M. F., Sun, D. M., Shao, H. F., Li, C. B., & Teng, Y. C. (2016). Poor sleep in middle-aged women is not associated with menopause per se. *Brazilian Journal of Medical and Biological Research.*, *49*(1).

van de Straat, V., & Bracke, P. (2015). How well does Europe sleep? A cross-national study of sleep problems in European older adults. *International Journal of Public Health*, *60*(6), 643–650.

Vézina-Im, L. A., Moreno, J. P., Thompson, D., Nicklas, T. A., & Baranowski, T. (2017). Individual, social and environmental determinants of sleep among women: Protocol for a systematic review and meta-analysis. *BMJ Open*, *7*(6), e016592. Available from: http://bmjopen.bmj.com/lookup/doi/10.1136/bmjopen-2017-016592.

Watson, N. F., Badr, M. S., Belenky, G., Bliwise, D. L., Buxton, O. M., Buysse, D., et al. (2015). Recommended amount of sleep for a healthy adult: A joint consensus statement of the American Academy of Sleep Medicine and Sleep Research Society. *Journal of Clinical Sleep Medicine. American Academy of Sleep Medicine*, 591–592.

Williams, N. J., Robbins, R., Rapoport, D., Allegrante, J. P., Cohall, A., Ogedgebe, G., et al. (2016). Tailored approach to sleep health education (TASHE): Study protocol for a web-based randomized controlled trial. *Trials*, *17*(1).

Zaharias, G. (2018). What is narrative-based medicine? *Canadian Family Physician*, 64.

CHAPTER 7

"Digital neurotic library"—A contribution to health literacy

Berta Maria Jesus Augusto, Carlos Manuel Santos Fernandes, and Sérgio Filipe Silva Abrunheiro
The Coimbra Hospital and University Centre, Coimbra, Portugal

In recent years, we have witnessed a paradigm shift regarding the citizen's position in health systems. It is no longer intended that he/she be a mere passive subject of care but rather currently advocating his/her involvement and co-responsibility throughout his/her life cycle in clinical decision-making that allows him/her to make informed decisions in favor of his/her health. This vision is reflected in a set of strategic documents on health in Portugal, such as the National Health Plan 2021–2030, the National Programme of Education for Health, Literacy and Self-Care, the National Plan for Patient Safety 2021–2026, and the Manual on Policies and Strategies for Quality of Health Care, among others. If we want to have a society with higher levels of health literacy, we have to think about the whole life cycle of the person, via a multi-sectoral approach, involving all citizens, where health policies must be integrated and meet people's real needs. The approach should also be multidimensional, considering the following dimensions: individual level, the community, health systems and policies, and multi-literacy (Arriaga, 2019).

It is recognized that citizens with higher levels of health literacy can better navigate the health system, better manage their health and their care plan, and better evaluate the care provided. In this way, they will use health services less frequently, namely emergency services, and will also be able to use them in a more rational way, contributing to their sustainability. Thus, it is necessary to strengthen the strategies and means to support health literacy, training, and empowerment of citizens, with a view to promoting their health and preventing disease, promoting their well-being, and increasing the effectiveness and efficiency of health systems (da Saúde, 2019).

Between 2020 and 2021, the Directorate-General for Health conducted an assessment of the Health Literacy of the Portuguese population, given the recognized importance of knowing this data for the process of health literacy promotion. The assessment is part of the Health Literacy Survey 2019 organized by the M-POHL consortium, which is taking place in 15 Member States of the European Region of the World Health

Active Learning for Digital Transformation in Healthcare Education, Training and Research
https://doi.org/10.1016/B978-0-443-15248-1.00018-7

Organization. The results found for the Health Literacy categories show a higher proportion of participants with high levels of Health Literacy (categories of sufficient and excellent) than with low levels of Health Literacy (categories of problematic or inadequate). The majority of people, namely 65%, were classified as having a sufficient level of health literacy and 5% as having an excellent level. In contrast, 7.5% of the people in the sample of this study were classified as having an inadequate level and 22% were classified as having a problematic level of health literacy. These results show a slight improvement compared to previous studies in this area but reveal the need to continue to invest in the promotion of health literacy (da Saúde, 2021).

Health literacy is defined as "the extent to which individuals have the ability to obtain, process and understand basic health information to use services and make appropriate health decisions" (WHO, 2020 (cited by DGS, 2019:3)), and it is essential for health professionals to be facilitators in the development of citizens' skills that promote access to, understanding of, and use of health information. For this to happen, health professionals, in their different contexts of intervention and interaction (individual and collective), should seek to activate health literacy in their contact with the person, using assertive, clear, and positive language, mitigating the risk of miscommunication, involving the person in the whole care process, and assuming themselves as a credible source of information, always validating the understanding and use of that information. They should also adjust measures to make the navigability of the health system clear and easy to understand, as well as reinforce people's efforts to adopt a healthy lifestyle (Arriaga, 2019). Studies point out health professionals as the vehicle for health information that the population seeks the most (Pedro, 2019).

The use of various pedagogical resources such as audiovisual media and interactive tools has been shown to facilitate the transmission of information between health professionals and patients, facilitating understanding and therapeutic adherence (Brito, 2020).

Considering the facts previously mentioned and the preponderance currently attributed to digital communication supports for the promotion of health literacy, the present "Digital Neuroteca" project was created. It originated from the needs felt by health professionals in the promotion of health literacy in the inpatient and outpatient service units of the Coimbra Hospital and University Centre, specifically the Inpatient Unit A and Day Hospital of the Neurology Service.

Neurological disease, which often generates dependence in self-care, requires changes in the person's lifestyle and it is essential to acquire a set of skills that allow them to carry out an effective self-management of their disease.

A new health condition requires a process of adaptation by the person, with the need to incorporate a set of changes in his/her life. This process includes the awareness of these changes and their implications in the new health condition, as well as the need to develop a set of skills so that this adaptation process is effective.

The understanding that previous behaviors may have been wrong and that a different behavior will facilitate the adaptation to the new health condition is a determinant for the person's involvement in the therapeutic process, and the health professional plays a key role in motivating this to happen. He/she should consider all the health determinants that may facilitate or hinder this process, enhancing or reducing them respectively. We refer to external determinants (education, socioeconomic status, community conditions) and internal determinants (meanings, beliefs and attitudes, level of preparation/knowledge).

As a result, it is necessary to empower the person as a facilitating intervention in the adaptation process, using good health literacy practices as a resource.

Based on these assumptions, the health professionals of this service felt the need to have credible and easily accessible information which could serve as a support to the empowerment process of the neurological patient and/or his/her informal caregiver.

Therefore, in the last quarter of 2019, it was decided to build a digital repository with a selection of content that is clear, credible, and targeted to the needs of the reference population of this service, given the amount and diversity of information available on the internet and its frequent lack of credibility.

In the genesis of the "Digital Neuroteca," several meetings of the multidisciplinary team were held, using the brainstorming technique, which aimed above all at the involvement and active participation of the entire team in the project, identifying the contents to be incorporated into the repository, defining the methodology of use based on good health literacy practice and defining the project's evaluation indicators.

Following the contributions provided by the various participants in the meetings, the leadership team systematized the entire project methodology based on a specific procedure, which it disclosed to the team. At the same time, the construction of the repository began, integrating various pedagogical resources such as videos, images, e-books, infographics, manuals, pamphlets, and flyers. In this selection, there was a concern about choosing contents with simpler and clearer language, more appealing and objective that could be understood even by citizens with lower levels of health literacy. Care was also taken to carry out pre-tests, involving patients and informal caregivers, so that they could validate the clarity of these contents. We also included hyperlinks to credible websites such as those of patients' and caregivers' associations and other community resources that may complement the information transmitted during hospitalization/outpatient treatment and support the neurological patient and/or informal caregiver in their process of adaptation to the new life situation.

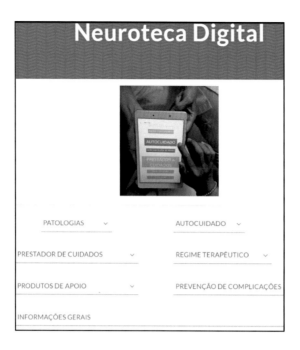

The selected contents were grouped into a set of categories, namely, Pathologies; Self-care; Therapeutic Regime; Support Products; Prevention of Complications; and Caregiver and General Information, as suggested by the health professionals that make up the Service's team. More specifically, information is available on: the various neurological diseases that are most prevalent in the service (stroke, dementias, multiple sclerosis, amyotrophic lateral sclerosis, etc.); adaptive strategies and support products for self-care (hygiene, dressing and undressing, eating, drinking, walking, positioning, transferring, etc.); therapeutic regime (healthy lifestyles, safe use of medicines, strategies for adherence to the therapeutic regime, etc.); prevention of complications (prevention of pressure ulcers, falling, aspiration in dysphagia situation, ankylosis, infection, etc.); information for the informal caregiver (care techniques, available support and resources, promotion of their well-being, accessibility to support products such as articulated beds, wheelchairs, pressure relief devices, transfer devices, self-care devices, etc.); and miscellaneous general information (living will, status of the informal caregiver, citizens' rights and duties, contacts of patients' associations, navigability in the health system, accessibility to health services, digital support apps, etc.).

This digital repository is accessible to all health professionals of the service who were previously trained in its use. In these training moments, in addition to training professionals for the use of the tool, the training focused on the most effective communication strategies in the therapeutic relationship with the person, on the use of the ACP model (assertiveness, clarity, and positivity), and on the need to address the dimensions of access, understanding, and use of health information. The information content is presented using tablets acquired for this purpose and made available by health professionals (preferably the reference professionals), who remain close to the patient and/or informal caregiver to clarify any doubts and validate their understanding.

The information is provided by the health professional, not randomly, but after a selection based on the assessment of the person's individual needs and characteristics. The information is transmitted in a piecemeal manner (chunk and check), with the intentional use of content repetition and using assertive, clear, and positive communication— ACP model (Almeida, 2019).

The fact that the person has access to this information in the inpatient context also allows them to validate the understanding of the information transmitted, for example, through the **teach-back** technique, giving them the opportunity to clarify misinterpretations and synthesize the contents considered most relevant.

To check whether the person knows how to use the information transmitted, health professionals put them in hypothetical situations related to their health situation that call for the mobilization of the knowledge obtained in order to solve them.

Where appropriate, or at the request of patients/informal carers, select content is made available in their email.

The contents hosted in this repository are regularly updated, due to the permanent dissemination of material of interest in this area, a process to which the entire multidisciplinary team contributes, centralized in the promotion team. At this stage of the project, there are already several materials, such as pamphlets, infographics, and videos, which integrate the "Digital Neurotic Library" prepared by the service's professionals, based on good health literacy practices.

For the implementation of the project, eight training sessions were held, with a total of 24 h, on the use of good health literacy practices and the use of the digital repository, addressed to all health professionals in the service. A questionnaire supplied to the professionals before and after the training session allowed us to conclude that they had a limited understanding of the concept of health literacy and were unaware of a set of strategies to promote health literacy.

It should be noted that the health professionals expressed their satisfaction with the training, demonstrating a desire to change their behavior, with a view to better results in this area.

In order to assess the satisfaction with this project, opinion surveys are applied twice a year to health professionals and patients and/or informal caregivers who are the target of training. Health professionals report satisfaction with the use of this resource due to the availability and easy access to information on multiple topics appropriate to the needs of the neurological patient, helping to clarify and standardize the message for the patient. Patients and/or informal caregivers reported equal satisfaction with the diversity and clarity of the contents made available and their adequacy to their needs. Their expressions included: "That video helped me understand the harmful effects of smoking, …".

Since the implementation of the project until the first semester of 2022, this repository has had 2166 accesses. The "top 10" categories and subcategories accessed are presented in Table 1.

In conclusion, we can say that the Digital Neurotic Library has proved to be very useful in promoting health literacy among the patient, the informal caregiver, and the health professionals themselves. These findings are in line with the scientific evidence, by pointing out the diversity of communicational strategies as a promoter of health literacy. The intention is to soon integrate primary health care into this project, which will allow continuity in the provision of contents as well as in their dissemination, enhancing the evaluation of the use of the information provided and their integration into the management of their new health condition.

Table 1 "Top 10" categories and subcategories accessed.

Category/subcategory
Multiple Sclerosis
Take care
Dietary regime
Support products
Medicinal regime
General Information
Caregiver Support
Falls
Dementia
Walking with a walking aid

References

Almeida, C. V. (2019). Modelo de comunicação em saúde ACP: As competências de comunicação no cerne de uma literacia em saúde transversal, holística e prática. In C. Lopes, & C. V. Almeida (Eds.), *Literacia em saúde na prática* (pp. 43–52). Lisboa: Edições ISPA. http://loja.ispa.pt/produto/literacia-em-saude-na-pratica.

Arriaga, M. T. (2019). Prefácio. Capacitação dos profissionais de saúde para uma melhor literacia em saúde do cidadão. In I. C. Lopes, & C. V. Almeida (Eds.), *Literacia em saúde na prática* (pp. 11–15). Lisboa: Edições ISPA. http://loja.ispa.pt/produto/literacia-em-saude-na-pratica.

Brito, D. V. (2020). O design também salva vidas - o poder escondido da literacia visual. In C. V. Almeida (Ed.), *Literacia em saúde, um desafio emergente – contributos para a mudança de comportamento* (pp. 47–51). Coimbra: Gabinete de Comunicação, Informação e Relações Públicas do Centro Hospitalar e Universitário de Coimbra. https://www.chuc.min-saude.pt/media/Literacia_Saude/Literacia_em_Saude_-_Coletanea_de_Comunicacoes.pdf.

da Saúde, D.-G. (Ed.). (2019). *Manual de boas práticas literacia em saúde: Capacitação dos profissionais de saúde*. Lisboa: DGS.

da Saúde, D.-G. (2021). *Níveis de literacia em saúde*. Lisboa, Portugal: DGS.

Pedro, A. (2019). *Literacia em saúde na doença crónica*. https://www.saudequeconta.org/wp-content/uploads/2020/01/Literacia-em-saude_doenca-cronica.pdf.

World Health Organization. (2020). *Why health literacy is important*. https://www.euro.who.int/en/health-topics/disease-prevention/health-literacy/why-healthliteracy-is-importante.

CHAPTER 8

Young people and health literacy: The digital influence

Cristina Vaz de Almeida[a,b], Diogo Franco Santos[c,d,e], and Patricia Martins[f,g]
[a]Instituto Superior de Ciências Sociais e Políticas (ISCSP), Lisbon, Portugal
[b]Portuguese Health Literacy Society (SPLS), Lisbon, Portugal
[c]General and Family Medicine Specialty Training
[d]Medicine at NOVA Medical School
[e]Health Literacy
[f]Portuguese Association for the Promotion of Public Health (APPSP)
[g]Regional Administration of Health of Lisbon and Tagus Valley (ARSLVT) to exercise functions in the Health Unit Public Arnaldo Sampaio of the Group of Health Centers (ACES) Arco Ribeirinho

1 Background

Health literacy (HL) is generally defined as "the degree to which individuals have the capacity to obtain, process and understand the basic health information and services needed to make appropriate health decisions" (World Health Organization [WHO], 1998). To invest in the promotion of HL among the younger generations is to invest in the improvement of the health and well-being of future generations. In addition to the daily management of their own health, young people will also be the caregivers of the future, not only regarding their children but also taking into account their parents and other relatives (Almeida, 2022). Digital HL refers to the ability of individuals to search, find, understand, and evaluate health information from electronic resources and apply this knowledge to solve a health problem (Norman & Skinner, 2006). Today's young people were born and are growing up in a technology-dominated world, navigating the internet and its shared consumption platforms in a very intuitive, fast, and logical way. At the same time, the growing risk of misinformation and fake news has been increasing, due to the frequent absence of "filters" regarding the suitability and relevance of the information available online. Stellefson et al. (2011) also emphasize that, although the current generation of young people has access to a wealth of health information on the internet, access alone does not guarantee that they are specialized in the search for health information. Thus, in addition to the problem of low HL, there is also the problem of low digital HL, given the difficulty that these young people may experience in understanding and assessing health information from electronic resources, as well as in applying this knowledge to solve a health problem (Almeida, 2022). It is undeniable, however, that the amount of information to which young people have access in their daily lives has never been as large as it is today. However, a study conducted in 2021 in Portugal (National School of Public Health—NOVA University Lisbon) revealed that 44% of

Active Learning for Digital Transformation in Healthcare Education, Training and Research
https://doi.org/10.1016/B978-0-443-15248-1.00007-2

Portuguese university students have an inadequate or problematic level of HL, a knowledge deficit that seems to be influenced by the socioeconomic background of their families. This study is a reminder that although the current generation of young people is the most academically qualified ever, their HL levels are still far from ideal (Pedro, 2022). Given the fact that we live in a world dominated by social networks and online platforms of shared consumption, it is important to reflect on strategies to improve the levels of HL of today's young people (Generations Y, Z, and Alpha), seeking appealing, interactive, and dynamic ways to reach them through digital media, widely used and disseminated globally.

2 Methods

A narrative review of Portuguese and international scientific publications was conducted using the MeSH terms [health literacy], [digital], [young people], [education], and [social media]. The inclusion criteria of the publications were related to their relevance to the topic of this article, which made the search extensive, as opposed to a systematic review (Bryman, 2012). Priority was given to articles published in the last decade, but some older articles that the authors considered relevant were also included.

3 Discussion

Ghaddar, Valerio, Garcia, and Hansen (2012) conducted an online survey with a cross-sectional random sample of high school students in South Texas, assessing their HL using the eHEALS and New Vital Sign scales, and found that among the 261 students who completed the survey, HL was positively associated with self-efficacy and online health information seeking. The authors concluded that exposure to a credible source of online health information was associated with higher levels of HL and thus recommended the incorporation of such information into the school health education curriculum to promote HL. Also according to the same authors—and adding to our view on the subject—, access, understanding, use of health information and resources, and promotion of HL among teenagers are essential for the following reasons: (1) Teenagers are developing health behaviors and habits throughout their lives, and the skills they have in HL may lead to more informed lifestyles in seeking health care; (2) teenagers are future independent users of the healthcare system, and young adults having more HL may contribute to a reduction in less favorable health outcomes; (3) teenagers are increasingly enjoying and accessing online health services as health systems have also been evolving in terms of web-based services, which have been greatly boosted by the recent COVID-19 pandemic; (4) the few studies investigating adolescent literacy and HL have shown that low literacy and low HL are associated with risk behaviors, including tobacco use, deviant behavior, and obesity; and (5) increased HL and digital HL empower young people to make more informed and responsible decisions about their health and that of their peers (family, friends and others).

In a recent study by Dadaczynski et al. (2021), search engines, news portals, and public agency websites were the most used by students as sources to search for information about COVID-19 and related topics. Female students used social media and health portals more frequently, while male students preferentially resorted to Wikipedia and other web-based encyclopedias, as well as YouTube. The use of social networks was associated with a low ability to critically evaluate information, while the opposite was observed for the use of public websites. Among the conclusions of Dadaczynski et al.'s research (2021), it is highlighted that, although digital HL is well developed in university students, many of them still have difficulties in assessing the information they find online, so it is essential to empower them in this sense, using personalized interventions, as well as to improve the quality of health information available on the internet.

According to studies carried out in Portugal (including the above-mentioned one by the National School of Public Health—NOVA University Lisbon), there is a pressing need to develop ecosystems that promote health and well-being, through the implementation of HL programs for children and young people, in order to empower them to make informed and enlightened daily decisions. Education and teaching establishments are, by excellence, a privileged environment for the promotion and education of health. Several projects and programs have already been implemented in Portuguese schools, as is the case of the National School Health Program (General Directorate of Health, 2005), the Network of Health Promoting Schools (SHE, 2014–2022), or the regime of application of sexual education in a school environment, established through Law no. 60/2009, of August 6th (Official State Gazette, 2009). Alongside these, other smaller but equally relevant projects have also been developed in Portugal, as is the case of two projects developed by researchers Ana Pires and Ana Coelho: the PEN (Period Empowerment Network) (2022) to demystify menstruation and the YHL (Youth Health Literacy) (2022), which aims to contribute to improving the HL of young people, particularly through the production of manuals for workers and youth organizations.

Check out the European Network of Health Promoting Schools: https://www.schoolsforhealth.org/

The European project ySKILLS (Youth Skills), which started in 2021 and is set to end in 2023, wants to know what the digital skills of European adolescents are and how they evolve (van Laar, van Deursen, Helsper, & Schneider, 2022). In 2021, 2022, and 2023, the same young people from six countries (Germany, Estonia, Finland, Italy, Poland, and Portugal) would answer the same questions about digital access, uses, and skills (van Laar et al., 2022). Researchers Ponte, Batista, and Baptista (2022), who were responsible for a Portuguese adaptation of this study, obtained 1017 answers to the survey from teenagers aged between 12 and 17 years. Regarding daily online activities, 89%

communicate with friends, 81% listen to music, 64% play games, 24% search for news, 22% look for new friends, 21% search for information on physical health and treatments, 18% search for information on mental health, 41% use the internet to learn new things, and 33% use it to practice something they are learning. During a school day, they use the internet for an average of 4 hours, with 41% reporting that they do not know how to check whether the information they found online is true and 35% not knowing how to assess whether a website is trustworthy (Ponte et al., 2022).

> **See here the 2021 results from the ySKILLS 2021 questionnaire reports from the participating countries: https://yskills.eu/publications/**

3.1 Participation and inclusion of young people

An active participation and inclusion of young people in the construction of HL programs (whether they are digital or not) seems essential to their success (Table 1).

The role of parents, teachers, health professionals, and the rest of the community network in which young people are involved is fundamental for the planning, implementation, and monitoring of these measures. They should be involved in the process of creating HL programs from the beginning of their design, taking into account their ideas and needs, and with a view to building information that responds, effectively and by the most appropriate means, to their doubts and concerns (Almeida, 2022).

According to UNESCO (2022), young people are encouraged to take action, from conception to implementation and follow-up, in their communities, through the scaling up of youth-led initiatives, and on the policy agenda, through mainstreaming their concerns and issues. To this end, UNESCO also encourages Youth Spaces, which aim to empower young people, foster and support their action, promote partnerships, and ensure their recognition and visibility.

Table 1 Measures to promote the participation and inclusion of young people in HL programs.

- Listening to young people's interests and motivations about their health;
- Understanding the content of social media and shared consumption platforms;
- Making use of audiovisual media common among young people (e.g., videos, *stories*);
- Taking into account the generational particularities of young people, such as the concern for sustainability and environmental protection or the awareness of LGBTQIA+ rights;
- Promoting peer education;
- Using the arts as a strategy for health promotion and disease prevention;
- Considering the importance of teleconsultation and the use of mobile phones for sending messages that can be an aid to communication and dissemination of information;
- Considering the creation of a subject such as 'Health' or 'Health Literacy' at an early stage of schooling (e.g., primary school);
- Involving young people in school strategies and policies leading to healthier lifestyles and a greater understanding of risk factors.

3.2 Health literacy in educational institutions

Schools and universities are privileged places for the implementation of HL programs, due to their natural proximity to the younger population. It is essential to integrate HL into the curricula throughout the academic pathway of children and young people, both at different educational levels and in different scientific areas (not only in health courses) (Almeida, 2022). As previously mentioned, these programs should privilege the active participation of children and young people at all stages, from planning to the evaluation of the implementation of the projects/programs; listening to their needs, motivations, and beliefs; and integrating their strategies and peer education—thus valuing their participation, the development of socioemotional skills, and the involvement of the whole school community and society. It is also important to mention that, prior to building the HL programs, it is essential to conduct a survey of training needs in each region/county/city, in order to better meet the young population in all their social, economic, and geographic context, thus bridging the barriers related to access, understanding, and use of health information.

In the context of the webinar hosted by the Portuguese Society for Health Literacy in May 2022 on "Young People and Health Literacy: What Challenges?", the speakers, Dr. Diogo Franco Santos and Nurse Patrícia Martins, highlighted the "need for the existence of school programmes, in partnership with health institutions, which aim to promote HL skills". They also mentioned the relevance of "the creation of an environment of safety and well-being, by the entire school community and the groups to which the young people belong to" as an HL-promoting measure (Sociedade Portuguesa de Literacia em Saúde & Miligrama - Comunicação em Saúde, n.d.).

> **Webinar "Young People and Health Literacy: What Challenges?"**
> https://www.youtube.com/watch?v=OwgagU6-NmM&ab_
> channel=SociedadePortuguesadeLiteraciaemSa%C3%BAde

It is also important to reflect on the evaluation of the implementation of health education programs implemented in schools and the low levels of HL among young people. The formal curriculum with more time allocated to health education and the training and capacity building of professionals in the promotion of health education in school settings are two necessary improvements.

3.3 The problem with "NEET" young people

A pertinent reflection is that it may not be possible to reach young people through educational institutions and in some cases not even through their jobs. This is where the concept of NEET ("Neither in Employment nor in Education or Training") comes in. According to the OECD (2022), these are young people who are neither in employment nor in formal education or training institutions.

NEETs are often the result of an anonymous and polarized branching out of people in communities without large support networks, as well as the lack of specific policies for preventive integration into occupational processes, which causes these young people to end up dropping out of the educational/learning system. Usually, after several failures at school and after entering and leaving precarious jobs, these young people are left "hanging" in a system that gives them the acronym of NEET, a term that, in itself, may have a negative connotation and may be stigmatizing. In this context, the book *Em Nome Próprio* (Almeida, 2017) gave voice to real and self-told experiences of a group of 50 young people from a school in Lisbon, as well as to their surrounding community. These manifestations of wills, beliefs, and assessment of the barriers behind these NEETs "help realize that they are also the result of a certain context, which runs alongside socio-economic hardship, housing contexts, opportunities for improvement, lack of parenting skills, also being an opportunity to make better public policies, with innovative strategies, able to see adolescents as actors of their own history" (2017, p. 7).

The promotion of HL among young NEETs should be integrated into community intervention programs, and within the framework of the programs and community dynamics, health professionals should promote the HL skills of local stakeholders as well.

3.4 Parental skills in the use of digital tools

As the younger generations—especially Generation Z and Generation Alpha—are digital natives, with new ways of relating, learning, and experiencing the world, it is important to adopt communication strategies with which these young people identify within social networks and other communication channels used by them, thus improving access to health information. The increase in their digital skills requires the development of parental digital skills, particularly in the use of tools that allow for the monitoring and supervision of contents, which, in some circumstances, put their health and lives in danger, as seen with several online challenges, such as the "Blue Whale Challenge", "Momo", "Goofy Man", "Frozen Honey", "Magnet", and, more recently, the "Blackout Challenge". The latter started on the social network TikTok and became popular among young people and was highly publicized for causing the death of a 12-year-old British boy, Archie Battersbee (Sky, 2022). These examples of online challenges, with possible physical and psychological consequences for young people, force us to rethink some practices and redouble the attention of parents and legal guardians (who should also acquire skills in digital HL), in addition to reinforcing the importance of intervention with the children and young people themselves.

Taking again the example of NEET young people, a preventive factor in this area also seems to be the development of parenting skills (Almeida, 2017), thus creating a safer and more positive environment that allows for a more effective understanding of health messages.

Learn more about cybersecurity at: https://www.seguranet.pt/

3.5 The role of health professionals

All health professionals are important for the promotion of greater HL among young people and they should also be prepared to understand this specific framework, equipping themselves with HL tools capable of supporting them in this task. Although any health professional can (and should) instill notions of HL in young people, we highlight in this context the importance of family doctors and their skills, which imply greater proximity and a life-cycle approach. The longitudinal monitoring performed by these professionals, as well as the provision of care in the community and the holistic approach, taking into account one's biopsychosocial, cultural, and existential dimensions (WONCA, 2002), allow for a more targeted action toward the whole person, as well as a greater knowledge of the family and community resources. Thus, the family doctor occupies a privileged position as far as promoting HL is concerned, with the potential to encourage other health professionals to follow the same steps, working together to promote health and prevent disease.

It is also important to remember that the promotion of HL also takes place outside of the physical space of the consultation itself, through its complementary communication channels, which are intuitively used by young people - as in the case of email and tele-consultation, which provide a certain comfort in interpersonal contact nowadays, and whose popularity rose after the onset of the COVID-19 pandemic. During the pandemic, Vaz de Almeida and colleagues conducted a study entitled "Digital Health in Times of Pandemic", with the purpose of knowing the respondents' opinions regarding their situation in relation to digital health and the means used. This study concluded that the respondents considered that the use of digital means in health was important (88.3%), which was already the case before the pandemic (72%), and that, during the pandemic, there was a 9% increase in the use of these services, namely teleconsultation, which increased from 37% to 46% (Almeida, Coelho, Martins, & Guarda, 2021). For this reason, it will be important to encourage investment in these contact networks, improving electronic circuits, and also through the training of professionals who use them on a daily basis and who so often face constraints in this context.

3.6 Challenges and concerns for young people

Adolescents do not navigate the health system as thoroughly as adults do, since they are generally healthy and, therefore, do not see the need for this research

(Ghaddar et al., 2012). Identifying the main challenges in the sense of engaging young people to make them informed citizens involves a critical and participatory reflection. It is important to take into account their profiles of action, life choices, trends, and concerns. In addition, and as mentioned above, alongside the large amount of information they have access to, young people are also exposed more than ever to fake news and sources of potential misinformation, which constitutes an obstacle to HL (Almeida, 2022).

The study (mentioned above) conducted by Stellefson et al. (2011) showed that students reported a reluctance to use interactive internet applications for the purpose of e-health communication with health professionals—a finding that, according to the authors, was found in the United States and also in a Finnish study. In this investigation, students did not mind using the internet to search for personal health information but were resistant to receiving individualized feedback about personal health concerns or problems through interaction with a qualified medical professional. This issue may be related to contextual web security issues affecting confidentiality and causing concern regarding data privacy, which also highlights the importance of understanding which online health information sources cause them to feel uneasy. Therefore, identifying, in the first instance, young people's barriers and concerns should be the first step in the construction of HL interventions aimed at them (Table 2).

Table 2 Considerations to take into account when improving young people's HL.

(1)	Allowing young people to speak about their full list of concerns (the classic example is during consultation with the health professional), using active listening techniques, avoiding early interruptions as they speak, and using open questions (when appropriate);
(2)	Providing an email or mobile phone contact to the young person as a way to clarify questions outside the physical space of the medical/nursing consultation;
(3)	Instilling notions of disease prevention (avoiding disease) and health promotion (having a healthy lifestyle);
(4)	Understanding young people's beliefs, intrinsic/extrinsic motivations, and concerns, taking into account their social and geographical context;
(5)	Replicating good HL practices, understanding what has worked and what can be improved, and above all, using HL techniques;
(6)	Using teleconsultation as an approach to digital media, as it will potentially be more appealing or desirable to young people;
(7)	Considering that young people who were born in a digital environment have different attitudes toward the world, which means that it is important to recognize the differences between Generations X, Y, and Alpha.

Source: Almeida, C. V. (2022). A Literacia em Saúde e os Jovens. Ponteditora. Available at: https://ponteditora.org/product/a-literacia-em-saude-e-os-jovens/.

3.7 The digital world and its particularities

The search for online information on health is, as already widely mentioned, an unavoidable tool when talking about young people. For this reason, in addition to the risk of misinformation and fake news inherent to digital platforms, there is another aspect that should be reflected on, which has to do with the fact that there are several immediate gratification factors (likes, stories, videos). These are usually organized by marketing strategies, which causes, according to the study by Kahneman (2011), young people to choose to make decisions using "system 1", i.e., acts by impulses (related to the reptilian brain), as opposed to "system 2", which is more thoughtful and reflective. In fact, human beings are always ready to take shortcuts and, therefore, there is a certain risk of bias in these decisions (Kahneman, 2011). Thus, and for the sake of developing their best skills (knowledge, attitudes and personal attributes), these issues related to immediate gratifications should be discussed to clarify the biases that lead to riskier decisions (such as smoking, consuming alcohol or drugs, or having an unhealthy diet), with worse health outcomes (Almeida, 2022).

A research study conducted within the scope of the Postgraduate Program "Health Literacy in Practice" (ISPA) (Bernardes, Santos, Taborda, & Barrambana, 2021), which analyzed a group of young people from Portugal and Brazil, belonging to Generation X, regarding their perceptions about high blood pressure, revealed that the format of choice for obtaining information on this disease were short 1-minute videos. In this line of thought, it may be relevant to mention that the networks where these videos are most used, such as TikTok or Instagram, should be subject to regular content reviews by qualified health organizations, in order to "screen" for digital material that may put at risk the physical and/or mental health of young people.

3.8 Communication at the heart of the matter

In the process of accompanying young people, and with regard to the health area more specifically, communication is undoubtedly a key element. As Watzlawick, Beavin, and Jackson (2000) also underlined, "everything is communication", and everything we do is communicating, whether by physical, digital, verbal, or nonverbal means. It is important to remember that HL can be built and worked on in many contexts, and not only in clinics or hospitals, since there are health organizations and groups, and not just health professionals. As an example, and in addition to websites, the presence and representation of health professionals in the various media used by young people, such as streaming platforms for films and series (e.g., Netflix, HBO), can also be a vehicle for transmitting correct and truthful messages about health, as a complement to and working in cooperation with the other more 'conventional' communication vehicles, such as medical

Table 3 Strategies for fostering young people's interest in HL.

- Knowing how to listen to their needs, concerns, and beliefs;
- Helping them to manage their emotions, which change from adolescence to adulthood;
- Making them see that the digital world can be user-friendly when it comes to accessing health information;
- Giving visibility to what they do and positively reinforcing their character and ideas;
- Giving feedback on their ideas and strategies ("hand them the microphone") and comment on the ideas, guiding them with appropriate suggestions;
- Taking into account the generational particularities (Generations X, Z, and Alpha), with regard to their interests, motivations, and concerns.

Source: Almeida, C. V. (2022). A Literacia em Saúde e os Jovens. Ponteditora. Available at: https://ponteditora.org/product/a-literacia-em-saude-e-os-jovens/.

consultations, physical formats (e.g., books, magazines and newspapers), or information leaflets. Engaging young people in HL requires a certain set of communication strategies (Table 3).

4 Conclusion

Higher levels of HL are undoubtedly associated with better health outcomes. The focus on the promotion of HL through digital media is a valuable tool to reach younger generations, who increasingly use these media earlier and more intelligently. The creation of multidisciplinary and intersectoral technical-scientific teams promoting HL programs, involving schools/faculties, social networks, media, and health institutions, with specific training in models, strategies, and techniques of HL and communication, is a necessary initiative. It is important to involve young people in all steps of the design of these strategies, as listening to their perspectives leads to the creation of public policies that better reflect their real needs. A focus on peer training is also a desirable goal to achieve in order to be able to plan, monitor, and critically evaluate the methods and results of HL programs. It is also important to ensure that HL reaches all young people, regardless of their social, economic, educational, or geographical background, and that there are mechanisms in place to verify the suitability of the health information available. Young people empowered with regard to their health will also improve the health of their peers and their community, in a network that is intended to be of interinfluence—and increasingly in the digital format, in line with technological developments and the needs of future generations.

References

Almeida, C. V. (2017). *Em Nome Próprio – Um guia para jovens e seus educadores, promotor de uma maior esperança.* Projeto BipZip. Projeto Alcântara Terra. Digiset.

Almeida, C. V. (2022). *A Literacia em Saúde e os Jovens.* Ponteditora. Available at: https://ponteditora.org/product/a-literacia-em-saude-e-os-jovens/.

Almeida, C. V., Coelho, I. D., Martins, P., & Guarda, L. (2021). *Saúde Digital em Tempos de Pandemia. Encontrar o sentido do espaço, comunicação e proximidade da saúde face-a-face, respeitando as diferenças.* Lisboa: APPSP. https://doi.org/10.5281/zenodo.4522750. Available at: http://appsp.org/site/assets/files/1223/2021-_relatorio_saude_digital_em_tempos_de_pandemia_-portugal_-_appsp.pdf.

Bernardes, C. P., Santos, D. F., Taborda, P., & Barrambana, S. P. (2021). Literacia em Saúde na Prevenção da Hipertensão Arterial em Jovens Adultos. In *II Encontro de Literacia em Saúde na Prática, 27 de novembro de 2021 (Ispa – Instituto Universitário).* Video of the presentation (2:44:01 to 3:08:17): https://www.facebook.com/watch/live/?ref=watch_permalink&v=780340506697697.

Bryman, A. (2012). *Social research methods* (4th ed.). Oxford University Press.

Dadaczynski, K., Okan, O., Messer, M., Leung, A. Y. M., Rosario, R., Darlington, E., et al. (2021). Digital health literacy and web-based information-seeking behaviors of university students in Germany during the COVID-19 pandemic: Cross-sectional survey study. *Journal of Medical Internet Research, 23*(1), e24097. https://doi.org/10.2196/24097.

Direção-Geral da Saúde [DGS]. (2005). *Programa Nacional de Saúde Escolar 2015.* Lisboa: DGS. Available at: https://observatorio-lisboa.eapn.pt/ficheiro/Programa-Nacional-de-Sa%C3%BAde-Escolar-2015.pdf.

Ghaddar, S. F., Valerio, M. A., Garcia, C. M., & Hansen, L. (2012). *Adolescent health literacy: The importance of credible sources for online health information.*

Kahneman, D. (2011). *Thinking, fast and slow.* New York: Farrar, Straus and Giroux.

Lei n.º 60/2009, de 6 de agosto, do Ministério da Educação. (2009). *Diário da República n.º 1515/2009, Série I, páginas 5097–5098.* https://files.dre.pt/1s/2009/08/15100/0509705098.pdf.

Norman, C. D., & Skinner, H. A. (2006). eHealth literacy: Essential skills for consumer health in a networked world. *Journal of Medical Internet Research, 8*(2), e9.

OECD. (2022). *Youth not in employment, education or training (NEET) (indicator).* https://doi.org/10.1787/72d1033a-en.

Pedro, A. R. (2022). Study: Health literacy in higher education: Challenges in Portugal. *APPSP, 18*, 2. http://appsp.org/site/assets/files/1231/newsletter_18.pdf.

Ponte, C., Batista, S., & Baptista, R. (2022). *Results of the 1st series of the ySKILLS questionnaire (2021) - Portugal.* KU Leuven, Leuven: ySKILLS. Available at: https://www.fcsh.unl.pt/static/documentos/informacao/ySkills_Relatório_Portugal.pdf.

SHE. (2014–2022). *Schools for health in Europe.* https://www.schoolsforhealth.org/.

Sky, B. (2022). *Archie, 12, tinha a morte marcada para as 11h00: o que todas as mães devem saber sobre ele e o TikTok.* August 3 CNN Portugal. https://cnnportugal.iol.pt/archie-battersbee/tik-tok/archie-12-anos-tinha-a-morte-marcada-para-as-11h00-o-que-todas-as-maes-e-os-pais-do-mundo-devem-saber-sobre-ele-e-o-tiktok/20220803/62ea3ec50cf2ea367d4832ca.

Sociedade Portuguesa de Literacia em Saúde & Miligrama - Comunicação em Saúde. (n.d.). *Webinar | Os Jovens e a Literacia em Saúde: Que Desafios?.* https://www.youtube.com/watch?v=OwgagU6-NmM&t=1s&ab_channel=SociedadePortuguesadeLiteraciaemSa%C3%BAde.

Stellefson, M., Hanik, B., Chaney, B., Channey, D., Tennant, B., & Chvarria, E. A. (2011). eHealth literacy among college students: A systematic review with implications for eHealth education. *The Journal of Medical Internet Research, 13*(4), e10.

UNESCO. (2022). *Youth.* https://www.unesco.org/en/youth.

van Laar, E., van Deursen, A. J. A. M., Helsper, E. J., & Schneider, L. S. (2022). *The youth digital skills performance tests: Report on the development of real-life tasks encompassing information navigation and processing, communication and interaction, and content creation and production skills.* KU Leuven, Leuven: ySKILLS.

Watzlawick, P., Beavin, J. H., & Jackson, D. D. (2000). *Pragmatics of human communication* (11th ed.). São Paulo: Cultrix.

World Health Organization [WHO]. (1998). *Health promotion glossary.* Geneva: WHO. https://www.who.int/healthpromotion/about/HPR%20Glossary%201998.pdf.

World Organization of Family Doctors (WONCA) Europe. (2002). *The European definition of general practice/family medicine (general practice/family medicine).* Barcelona: WONCA. https://www.woncaeurope.org/file/b662cccc-6ad6-4d34-a9a2-fd02d29fae5b/European%20Definition%20in%20Portuguese.pdf.

Further reading

Almeida, C. V., Lopes, C., Gonçalves, B., Nascimento, C. A., Bernardes, C. P., Marques, C., et al. (2021a). *Relatório: Storytelling: Pela voz e criatividade de profissionais das áreas da saúde*. Lisboa: ISPA – Instituto universitário.

Almeida, C. V., Lopes, C., Gonçalves, B., Nascimento, C. A., Bernardes, C. P., Marques, C., et al. (2021b). *Relatório: Vacinação, Estratégias de Comunicação e Literacia em Saúde Eficazes para População Jovem*. Lisboa: ISPA – Instituto universitário.

Atlântica – Instituto Universitário. (2022a). *PERIOD Empowerment Network [PEN]*. https://www.uatlantica.pt/pen/.

Atlântica – Instituto Universitário. (2022b). *Youth Health Literacy [YHL]*. https://www.uatlantica.pt/youth-health-literacy-yhl/.

CHAPTER 9

School-based interventions using media technologies to promote health behavior change and active learning about nutrition: A systematic literature review and meta-analysis

Sara Henriques[a], Manuel José Damásio[a], and Pedro Joel Rosa[b,c]
[a]Centre for Research in Applied Communication Culture and New Technologies, Lusófona University, Lisbon, Portugal
[b]Lusófona University, HEI-Lab: Digital Human-Environment Interaction Labs, Lisbon, Portugal
[c]Instituto Superior Manuel Teixeira Gomes (ISMAT), Portimão, Portugal

1 Introduction

Obesity is a serious public health problem, especially in children. The number of obese children has been increasing globally. Overweight and obese children are likely to stay obese into adulthood and more likely to develop health problems at a younger age (Oude Luttikhuis et al., 2009; Sahoo et al., 2015; WHO, 2009, 2021, 2022).

One of the main causes of becoming overweight is unhealthy nutrition, resulting in unbalanced energy consumption. Other factors include genetics, metabolism, sleep habits, psychological issues, environmental factors, and sedentary lifestyle (Sahoo et al., 2015). When speaking of dietary factors that contribute to weight gain, these include fast food, sugar-sweetened beverages, snack foods, and portion sizes (Nollen et al., 2014; Sahoo et al., 2015). On the other side, there are also specific foods that are considered to promote a healthier life, such as fruit, vegetables, brown bread, whole-grain cereals, and drinking water (Jones, Madden, & Wengreen, 2014; Pedersen, Grønhøj, & Thøgersen, 2016; Rees, Bakhshi, Surujlal-Harry, Stasinopoulos, & Baker, 2010).

Overweight and obesity conditions are associated with serious health problems, such as noncommunicable diseases (NCD)—cardiovascular diseases, musculoskeletal disorders, type 2 diabetes, some types of cancers, respiratory diseases, hypertension, hyperlipidemia, liver, and renal diseases (Bandini, Curtin, Hamad, Tybor, & Must, 2005; Sahoo et al., 2015; WHO, 2021, 2022). Noncommunicable diseases are one of the greatest challenges in health in the 21st century.

Active Learning for Digital Transformation in Healthcare Education, Training and Research
https://doi.org/10.1016/B978-0-443-15248-1.00014-X

Lifestyle plays a relevant role in noncommunicable diseases, as these are a result of risky behaviors and less healthy choices throughout life. Assuming that we all have control over our conduct, many of these risky behaviors can be preventable and modifiable. Promoting a willingness to change, to make health-enhancing choices, and to change established risky behaviors is a crucial matter that calls for research and for innovative and promising health promotion programs and interventions. The development of health interventions with an effective power to change behaviors represents a genuine public health challenge in current years.

Health promotion programs play a relevant role in creating healthy individuals, families, and communities. The goal of these programs is to promote health indicators and health behaviors and to foster informed decisions about health, by conducting planned and structured activities, using a specific setting, and focusing on a target audience (Fertman & Allenworth, 2010).

The school environment can work as an ideal setting when working with children (De Bourdeaudhuij et al., 2010; Whittemore, Jeon, & Grey, 2013). School-based interventions are some of the most used interventions to reach children, as well as offer promising results (WHO, 2009). These interventions have proliferated in the last decades (Whittemore et al., 2013); however, the cost of such interventions can be high, as well as the burden on educational staff. To answer to this problem, computer-tailored interventions and interventions using media technologies have emerged as a promising, cost-effective, and attractive strategy, being able to provide feedback and deliver content with relatively low cost and low effort while also promoting motivation and engagement (Whittemore et al., 2013; Yang et al., 2017). Technology and new media are encouraging tools for achieving health promotion goals as they often offer interaction with other people, peers, or experts; offer access to online information; and have the potential to deliver entertaining activities for health management (Street, Gold, & Manning, 2013). On the other side, children are confident and enthusiastic adopters of new media, usually sharing a positive perspective and being particularly tech savvy and tech engaged. However, few empirical studies have integrated mobile technologies, social interaction possibilities, and game-based designs addressing nutrition and health behaviors toward nutrition (Yang et al., 2017). Some studies have shown positive effects of health interventions on dietary habits by using new media as the main communication tool, such as online games, serious games, videogames (Banos, Cebolla, Oliver, Alcaniz, & Botella, 2013; Cullen, Liu, & Thompson, 2016; Jones et al., 2014; Joyner et al., 2017; Moore et al., 2009; Yang et al., 2017), mobile media (Nollen et al., 2014; Struempler, Parmer, Mastropietro, Arsiwalla, & Bubb, 2014), computer-based interventions (Brug, Campbell, & van Assema, 1999; Rees et al., 2010), and SMS interventions (Pedersen et al., 2016; Silva et al., 2015).

2 Method

2.1 Objectives

To analyze school-based promotion interventions using media technologies in nutrition with children aged 6 to 14. More particularly, this study aims to (1) synthesize the best available evidence on the structure, methodology, contents, impacts, and effectiveness of prevention programs using media technologies designed to promote healthy nutrition habits; (2) enumerate and understand the main theoretical frameworks and models of health behavior change used to guide interventions in this particular context; (3) identify main variables, moderators, and predictors of health behavior change to include in an intervention; (4) explore the main validated scientific instruments and tools used to assess health behavior change; and (5) explore the most used media technologies and their effectiveness on knowledge development and actual behavior practices.

2.2 Criteria for including studies in the review

This review included full papers of randomized control studies, experimental studies, and quasiexperimental and quasirandomized control studies on nutritional school-based interventions with children aged 6 to 14 years old, published in English and Portuguese from 2009 to 2019. Studies that measured outcomes immediately after and sometime after (days or months), such as follow-up measurements, were included.

Health interventions aimed at nonhealthy subjects or health interventions outside of a school setting were excluded. Studies including children aged less or more than this interval were included only if the mean age of the study sample was within this interval at baseline assessment. Studies that include other variables, such as exercise, were only included when they allowed measurable outcomes related only to nutrition. To conduct a meta-analysis, the following criteria were also complied with: report of minimal statistical data for the estimation of effect sizes. When there was missing data, the main author was contacted to provide additional data.

2.3 Search methods, terms, and classifications

Relevant studies were searched via scientific electronic databases and Internet search engines. The following databases indexed in EBSCO & ISI Web of Knowledge were searched: EBSCO E-Journals; Academic Search Complete; PsycINFO; Regional Business News; Newswires; Library, Information Science & Technology Abstracts; ERIC; Business Source Complete; Psychology and Behavioral Sciences Collection; Psyc-BOOKS; Cochrane Library CENTRAL; Scielo; and PubMed NCBI.

Further procedures were performed to ensure search breadth: (1) hand search of bibliographies of included studies; (2) hand search of websites of organizations; and (3) gray literature also included, such as those from search engines such as Google Scholar, conference proceedings, and thesis repositories.

Search terms were developed using relevant keywords. These terms were organized to produce a Boolean search phrase for each combination. The following table presents the keywords used in the search. Logical Boolean operators were used between columns ("AND") and between rows ("OR") to complete the search. The asterisk worked as a wildcard, words matched the search if they started with the word/letters preceding the asterisk (Table 1).

2.4 Data collection process and analysis

An initial search yielded 4024 results. Date range ($n = 2564$) and language were considered baseline criteria ($n = 2500$; English $= 2490$; Portuguese $= 9$). Additional records were identified through other searches ($n = 541$). The total sample included 3041 records.

Articles were imported to EndNote reference manager and screened independently with the support of Covidence software. Duplicates were removed (final sample $n = 2987$). This first round was performed based on the title and abstract reading. Those

Table 1 Keywords and Boolean search.

Nutrition	Intervention	Children	Media technologies
Eating	Health promotion	Child	ICT
Diet	Health intervention	Teen	Information technology*
Dietary	Health education	Teenager*	Communication technology*
Food	Health literacy	Youth	Media technology*
Food intake	Training program*	Adolescent*	Media
Food consumption	Gamification		Mobile phone
Nourishment	School program		Phone
Vegetable	School intervention		SMS
Fruit			Text message
			Flyer
			TV
			Television
			Radio
			PC
			Computer
			Online
			Tablet*
			iPad*
			Mobile app*
			Game
			Digital game
			Computer game
			Online game

that did not meet the eligibility criteria were excluded (sample $n=661$). Articles were screened against the eligibility criteria for type of study, type of participants, type of outcome measures, type of intervention, and type of setting (sample $n=48$). Afterward, each full-paper article was independently examined; 32 were excluded and 16 constituted the final sample. The following flowchart illustrates the process (adapted from PRISMA) (Fig. 1).

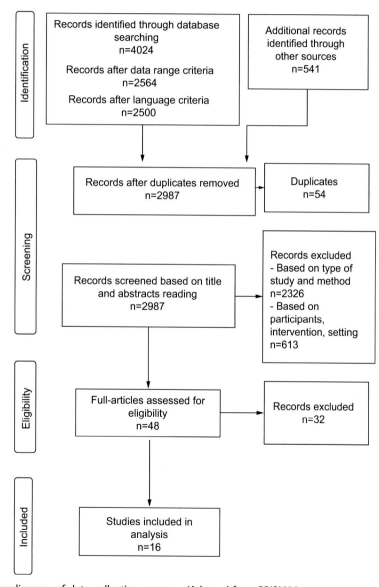

Fig. 1 Flow diagram of data collection process. *(Adapted from PRISMA.)*

3 Results

A comparative table was developed offering information on the type of study, participants, intervention, theoretical models, instruments, outcomes, ICT used, analysis, results, and conclusions (Annex 1).

3.1 Description of studies—Type of studies, location, and participants

A total of 16 records were identified and eligible for this review. Most studies were conducted in the United States (8), followed by Denmark (2), Portugal (2), Korea (1), Netherlands (1), Spain (1), and the United Kingdom (1).

Regarding the type of research design, nine studies are randomized controlled trials (RCT) (of these, two are cluster RCT). Five use a quasiexperimental research design, applying measure outcomes for treatment and control group; however, participants are not randomly assigned; usually, schools were the unit of analysis, mostly due to ethical and practical reasons. About two studies used a single-case experimental design, alternating treatment and no treatment days, using both pretests and posttests to compare dependent variable results. In both these cases, the whole school (all classes) was part of the intervention (collaborative gameplay) and a comparison was made alternating the target dietary behavior. The dependent variable was repeatedly measured across conditions (intervention days and nonintervention days). The target behavior was randomly selected, and students were notified daily. In these cases, the school served as treatment and its own control with repeated measurements across phases.

Participants were mostly male and female, except for two studies that only worked with female groups; in both cases, the intervention took place in girls-only schools and no particular reason was offered for that. As mentioned before, children were aged 6 to 14 years old. One-eighth of the studies included children aged between 6 and 8; half of the studies included children aged 9 to 11; and six studies included children aged 12 to 14. All studies included school-based interventions. Samples ranged from 49 to 2477 students, with a mean of 511 individuals (SD $= 624.347$).

3.2 Interventions

All studies focused on dietary practices, the main purpose being to promote healthy eating. Some 11 studies intended to promote healthy eating by increasing the intake of fruit and vegetables. Another eight studies included not only food behavior but also physical activity. One-quarter of the studies focused on general healthy eating and nutritional practices based on quantities and diversity. Nearly one-quarter focused on promoting nutritional knowledge. One study dealt with brown bread consumption, wholegrain cereals, fruit, and vegetables. Only two focused on decreasing sugar-sweetened beverages, sugar consumption, or snacks. One focused on daily water consumption.

The duration of the interventions varied greatly, from 2 weeks to 1 year, with an average duration of 12 weeks (SD = 11.214). Most of the studies used a baseline assessment and postassessment (after the intervention), and a quarter performed a follow-up (1 to 4 months after the intervention).

3.3 Theoretical background

For several studies, it was unclear if a theoretical model had been used to guide the intervention. Literature reviews focused mainly on explaining the problem of obesity, and some focused on the digital media technology that was used for the intervention. Regarding the studies that did express the use of theoretical background, the most common were the social cognitive theory and self-efficacy, the theory of planned behavior, gamification, serious games, game-based learning, collaborative learning, adoption theory, goal theory (goal setting strategy), experiential learning theory, self-care deficit theory, and implementation intention theory.

3.4 Media and communication technologies used

The most common media technology used in interventions were games, namely videogames, online games, and serious games (7 interventions); SMS (SMS-based monitoring and feedback system) (4 interventions); and mobile apps, Internet, computer-tailored leaflet, IPTV, and iPads (one intervention each). There were no interventions using more than one digital technology.

3.5 Outcome measures and instruments

The main outcome measures were based on dietary intake and nutritional practices. Most studies measured nutrition intake (fruit, vegetables, brown bread, water, sugar-sweetened beverages), nutrition knowledge, nutrition practices, anthropometric indices (blood pressure, height, weight, BMI), and psychological outcomes (self-efficacy, beliefs, cognitive and emotional factors).

The following list represents a comprehensive list of all outcomes measured: blood pressure, BMI, body fat, brown bread, calcium, carbohydrates, fiber, fruit, fruit and vegetables (FV), nutrition behavior, nutrition knowledge, protein, self-care practices, self-efficacy, snacks, sugar, sugar-sweetened beverages, total fat, vegetables, water, and wholegrain cereals.

The most common methods used to assess these outcomes were checklists, questionnaires, and surveys (11 studies). Around half of the studies used 24-h recalls (self-report of breakfast, lunch, dinner, snacks). Another half used anthropometry, height, weight, BMI, waist circumference, and blood pressure. Only two studies used plate waste measure and one study analyzed psychological outcomes.

There is a lack of uniformity in the instruments and measures used. Most of the studies developed their own tailored measures.; only a few used already validated measures. This reflects a need for the development, validation, and dissemination of more reliable measures in this area, able to evidence the impact of the interventions and gather data on the intended outcomes. This will allow for a more precise assessment and consistency between studies. The following table presents the list of instruments identified (Table 2).

3.6 Effect of interventions

A meta-analysis was conducted to aggregate the results of the 16 studies and analyze the combined efficacy. This aggregation is a procedure that weights the results of each study according to its precision (estimated based on dispersion).

A random effect model was chosen, as there is a low but significant degree of variability between the 16 studies ($Q=15.527$, df$=15$; $I^2=3.395$, $P<.001$). This procedure considers statistical heterogeneity and allows for the possibility that population parameter values vary, offering a wider confidence interval and a more conservative and reliable model that best represents variability (Higgins et al., 2019; Monteiro, 2010). Further exploration of the causes of heterogeneity is approached later in this analysis.

The following image (Fig. 2) shows the forest plot obtained with the random effect model. The Hedges' g for the 16 studies is 0.209, standard error of 0.042, with a confidence interval of 0.126 to 0.291. Data indicate that health nutritional interventions using ICT have a low to moderate significant effect on promoting healthier nutrition habits in children ($Z=4.950$; SE$=0.042$; $P<.001$). As there was more than one outcome

Table 2 Scientific instruments/measures.

Checklist: What's for lunch?	Struempler et al. (2014)
Nutrition knowledge questionnaire (28 items)	Moore et al. (2009)
Nutrition knowledge questionnaire for children (35 items)	Banos et al. (2013)
Psychological questionnaire (beliefs) (30 items for each food)	Rees et al. (2010)
Adolescent nutrition self-care questionnaire (50 items)	Moore et al. (2009)
Food frequency and physical activity questionnaire	Silva et al. (2015)
Food frequency questionnaire	Vereecken and Maes (2003)
Food frequency questionnaire	Ezendam, Brug, and Oenema (2012)
Family eating and activity habits questionnaire	Golden and Earp (2012)
Self-efficacy survey	Perry, Dewine, Duffy, and Vance (2008)

School-based interventions using ICT to promote healthy nutrition habits among children (6 to 14)

Study name	Outcome	Statistics for each study							Hedges's g and 95% CI
		Hedges's g	Standard error	Variance	Lower limit	Upper limit	Z-Value	p-Value	
Banos, et al., 2013	Combined	0,144	0,142	0,020	-0,134	0,421	1,014	0,310	
Baranowski et al, 2012	Combined	0,279	0,189	0,036	-0,092	0,649	1,475	0,140	
Bech-Larse & Gronhoj, 2013	Combined	0,153	0,132	0,017	-0,105	0,411	1,162	0,245	
Cullen et al., 2016	Combined	0,242	0,147	0,022	-0,047	0,531	1,644	0,100	
Enzedam, Brug & Oenema, 2012	Combined	0,185	0,071	0,005	0,045	0,325	2,593	0,010	
Fassnacht et al., 2015	Fruit & Vegetables	0,892	0,299	0,089	0,306	1,478	2,982	0,003	
Jones, Madden & Wengreen, 2014	Combined	0,174	0,063	0,004	0,050	0,299	2,748	0,006	
Joyner et al., 2017	Combined	0,301	0,120	0,014	0,066	0,537	2,509	0,012	
Moore et al., 2009	Combined	0,247	0,127	0,016	-0,001	0,495	1,954	0,051	
Nollen et al., 2014	Combined	0,276	0,278	0,078	-0,270	0,822	0,991	0,322	
Pedersen, Grnhoj & Thogensen, 2016	Combined	0,021	0,063	0,004	-0,103	0,145	0,330	0,741	
Rees& Baker, 2010	Combined	0,046	0,079	0,006	-0,109	0,201	0,578	0,563	
Sharma et al., 2015	Combined	0,173	0,206	0,043	-0,232	0,578	0,838	0,402	
Silva et al., 2015	Fruit & Vegetables	0,490	0,178	0,032	0,142	0,838	2,757	0,006	
Struempler et al., 2014	Combined	0,085	0,024	0,001	0,037	0,133	3,481	0,000	
Yang et al. 2017	Combined	0,476	0,074	0,005	0,332	0,620	6,458	0,000	
		0,209	0,042	0,002	0,126	0,291	4,950	0,000	

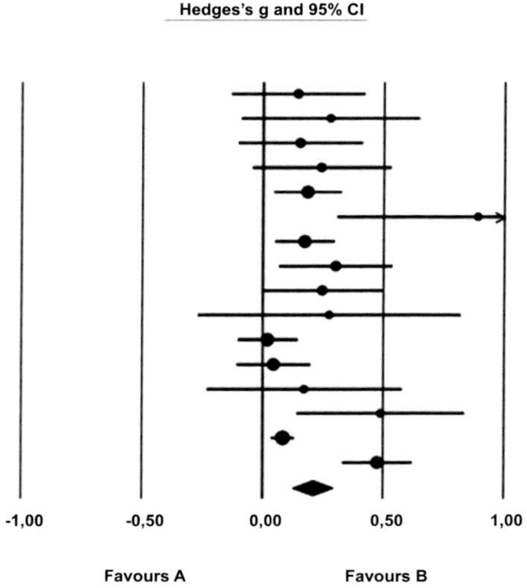

-1,00 -0,50 0,00 0,50 1,00

Favours A   Favours B

Meta-Analysis

Fig. 2 Forest plot.

measured in each study, the software computed a combined measure based on a mean of all outcomes per study, respecting the assumption of interdependence.

To test for the presence of publication biases, multiple methods were used. A funnel plot by precision (Fig. 3), allowing the visual analysis of systematic biases of publications (Fig. 4), shows a large concentration of studies around the average effect size, showing slightly more studies around the right side than the left side.

Considering the assumption that studies that indicate a significant difference have higher probability of being published in comparison with studies with nonsignificant or negative results, Duval and Tweedie's trim and fill method was used to estimate the number of missing studies on the left or right of the summary effect, suggesting five imputed studies. Egger's regression was also used, demonstrating that there is no evidence indicating a strong concern with publication bias ($P=.137$; CI: -1.018 to 4.354; $t=1.270$, df$=14$).

Further exploring the results, a search for the causes of heterogeneity and evaluation of outliers was conducted, as well as a subgroup analysis. We believe variability was found mostly due to methodological reasons (differences in research design and intervention) and clinical reasons (differences in participants, intervention, type of food/nutrition approached in the intervention). Measurements in human and social sciences have an inherent natural heterogeneity. Theoretically, the group of studies included in this analysis falls under a specific category of interventions and they have all undergone a straight inclusion criteria selection. However, these studies vary in terms of individuals' age, different media used (TV, games/videogames/collaborative games, digital games,

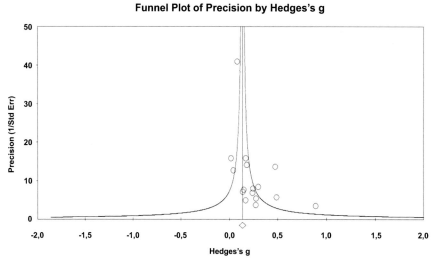

Fig. 3 Funnel plot by precision.

Schoolbased interventions using ICT to promote healthy nutrition habits among children (6 to 14)

Group by intervention	Study name	Outcome	Statistics for each study						
			Hedges'sg	Standard error	Variance	Lower limit	Upper limit	Z-Value	p-Value
Diet	Bech-Larsen & Gronhoj, 2013	Combined	0,153	0,132	0,017	-0,105	0,411	1,162	0,245
Diet	Cullen et al., 2016	Combined	0,242	0,147	0,022	-0,047	0,531	1,644	0,100
Diet	Jones, Madden, Wengreen, 2014	Combined	0,174	0,063	0,004	0,050	0,299	2,748	0,006
Diet	Joyner et al., 2017	Combined	0,301	0,120	0,014	0,066	0,537	2,509	0,012
Diet	Nollen et al., 2014	Combined	0,276	0,278	0,078	-0,270	0,822	0,991	0,322
Diet	Pedersen, Gronhoj & Thogersen, 2016	Combined	0,021	0,063	0,004	-0,103	0,145	0,330	0,741
Diet	Rees & Baker, 2010	Combined	0,046	0,079	0,006	-0,109	0,201	0,578	0,563
Diet	Struempler et al., 2014	Combined	0,085	0,024	0,001	0,037	0,133	3,481	0,000
Diet			0,119	0,040	0,002	0,041	0,198	2,996	0,003
Diet & exercise	Banos et al., 2013	Combined	0,144	0,142	0,020	-0,134	0,421	1,014	0,310
Diet & exercise	Baranowski et al., 2012	Combined	0,279	0,189	0,036	-0,092	0,649	1,475	0,140
Diet & exercise	Enzedam, Brug & Oenema, 2012	Combined	0,185	0,071	0,005	0,045	0,325	2,593	0,010
Diet & exercise	Fassnacht et al., 2015	Fruit & Vegetables	0,892	0,299	0,089	0,306	1,478	2,982	0,003
Diet & exercise	Moore et al., 2009	Combined	0,247	0,127	0,016	-0,001	0,495	1,954	0,051
Diet & exercise	Sharma et al., 2015	Combined	0,173	0,206	0,043	-0,232	0,578	0,838	0,402
Diet & exercise	Silva et al., 2015	Fruit & Vegetables	0,490	0,178	0,032	0,142	0,838	2,757	0,006
Diet & exercise	Yang et al., 2017	Combined	0,476	0,074	0,005	0,332	0,620	6,458	0,000
Diet & exercise			0,314	0,052	0,003	0,211	0,416	5,984	0,000

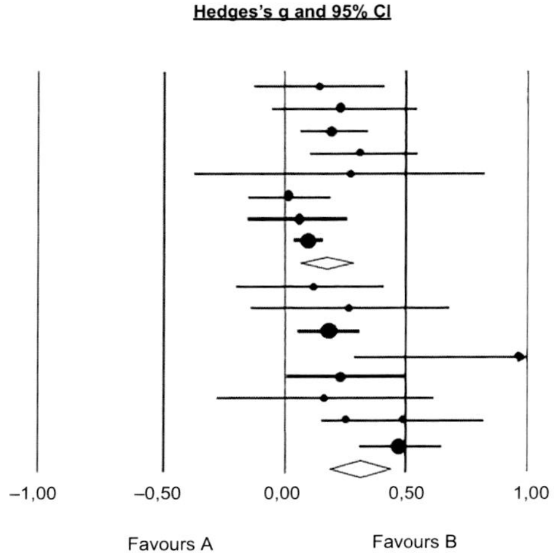

Hedges's g and 95% CI

-1,00   -0,50   0,00   0,50   1,00

Favours A   Favours B

Meta-Analysis

Fig. 4 Forest plot.

online/Internet, SMS/mobile, iPads/computers), duration of the interventions, addition of (or not) physical exercise to the intervention, and the type of food or nutrition approached in the intervention (promote healthy food or reduce unhealthy food consumption). A sensitivity analysis (to outliers) revealed low power to explain this variability.

3.7 Subgroup analysis for heterogeneity exploration

A subgroup analysis was considered the best method to explore and explain the causes of heterogeneity and draw valid conclusions (Higgins et al., 2019; Richardson, Garner, & Donegan, 2019). The subgroup analysis distinguished between studies where intervention focused only on nutrition (called "Diet" group) and studies where intervention focused on diet as well as some kind of physical exercise ("Diet&Exerc"). The Diet subgroup included 8 studies (3674 participants in the treatment group, 2806 participants in the control group) and the Diet&Exerc subgroup included 8 studies (1146 participants in the treatment group and 1207 in the control group). A random effect model was used to combine studies within each subgroup. A fixed effect model was used to combine subgroups. The study-to-study variance is assumed to be equal for subgroups, being computed within subgroups and pooled across subgroups. The following image presents the forest plot obtained.

For the Diet subgroup, the mean effect size in Hedges' g is 0.119, with a CI of 0.041 to 0.198, a Z-value of 2.996, and a corresponding P-value of $<.001$. This indicates that this type of intervention has a low but statistically significant impact. For the Diet&Exerc subgroup, Hedges's g is 0.314, with a CI of 0.211 to 0.416, a Z-value of 5.984, and a P-value of <0.001, indicating a moderate statistically significant impact. Enough studies were included in each subgroup, clearing any covariate distribution concerns.

The second group (Diet&Exerc) effect size indicates a higher impact, meaning that health interventions promoting nutrition habits would benefit from including physical exercise as well. The test for subgroup differences suggests that there is a statistically significant subgroup effect of the type of intervention used ($P=.003$). Regarding heterogeneity, results indicate that a subgroup analysis is the best method to analyze data, as results indicate a lower and nonsignificant heterogeneity within each subgroup (subgroup Diet: $Q=4.685$; df$=7$; $P=.330$; $I^2=0.748$; subgroup Diet&Exerc: $Q=10.326$; df$=7$; $P=.053$; $I^2=32.207$). This result indicates that physical exercise can potentially work as a moderator variable, strengthening the effect of health interventions in this area.

3.8 Further data exploration

Subgroup analysis based on outcomes and on media technology used within the interventions was performed to further explore the data. The goal was to understand whether the interventions have had a higher impact on a specific outcome, or whether a specific technology is more effective in producing the desirable outcomes of the health interventions.

The study sample measured a considerable list of outcomes and used a large list of media. Only those that were applied in more than two studies were selected for analysis, even though this still represents a very small sample, meaning that results should only be understood as indicative, needing additional confirmation, and thus, used with caution.

Results indicate that interventions using media to promote healthy nutrition among children (6 to 14) have a significant positive impact on the following:

- Fruit and vegetables intake ($g=0.471$; 95% CI $=0.257$ to 0.686; $Z=4.313$; $P<.001$)
- Vegetables intake ($g=0.287$; 95% CI $=0.125$ to 0.449; $Z=3.466$; $P<.001$)
- Fruit intake ($g=0.094$; 95% CI $=0.055$ to 0.134; $Z=4.651$; $P<.001$)
- Blood pressure ($g=0.505$; 95% CI $=-0.200$ to -0.809; $Z=3.249$; $P=.001$)
- BMI ($g=0.240$; 95% CI $=-0.123$ to -0.357; $Z=4.015$; $P<.001$).

A subgroup analysis focused on media technologies used to deliver the interventions indicates a greater effectiveness of the following media:

- Online games ($g=0.237$; 95% CI $=0.331$ to 0.705; $Z=5.427$; $P<.001$)
- SMS ($g=0.195$; 95% CI $=0.031$ to 0.358; $Z=2.337$; $P=.019$);
- Online lessons ($g=0.096$; 95% CI $=0.050$ to 0.141; $Z=4.135$; $P<.001$).

4 Conclusions

Childhood experiences shape adult life and have an impact on health status later in life (Langford et al., 2015). Consequently, health promotion interventions in childhood are of great relevance. Intervening to promote health behaviors, before risky behaviors are part of the daily life of the individual, is a fundamental strategy to promote a healthier society.

School-based health promotion interventions are an important tool to foster children's nourishment awareness and promote well-informed dietary choices.

Data from this study indicated that school-based nutrition interventions have a significant positive impact on promoting healthier nutrition among children. Moreover, results indicated that including physical exercise in nutrition interventions is beneficial and strengthens the effect of the interventions to have a stronger significant impact. Serious online games emerge as a promising tool for health interventions with children, followed by SMS and online lessons (e-learning). Fruit and vegetable intake group revealed the best results, which indicates that interventions were frequently able to promote (increase) the consumption of these foods in schools, with children.

The interventions had an average duration of 3 months and only a quarter performed a follow-up. A need for interventions that focus on promoting long-term effects is present. Results also indicated that it was not clear whether interventions have been developed under a particular theoretical framework, even though the literature suggests a higher effectiveness when studies apply theory to guide interventions (Henriques & Damasio, 2021; Cane, O'Connor, & Michie, 2012; Painter, Borba, Hynes, Mays, & Glanz, 2008). The most commonly used instruments to measure outcomes were questionnaires and surveys that were developed for the purpose of each intervention. A lack of validated commonly used instruments was noted.

The results of this literature review show that, although sharing a common goal (to promote healthier nutrition), interventions used different paths to achieve this goal. Nutrition is a very broad term. Some interventions focused on healthy eating (fruit, vegetables, brown bread, wholegrain cereals), while others focused on reducing unhealthy eating (sugar-sweetened beverages, snack, sugar), some on nutritional knowledge, and a few on shaping intentions and behaviors. Furthermore, interventions in this area act at different stages of the health behavior process, with some focusing on early stages (mainly promotion knowledge and motivation) and others focusing on later stages (fostering and measuring intentions and actual behaviors).

Adding to this complexity, nutrition is a very comprehensive concept, including different foods (fruits, vegetables, leguminous, meat, fish, pasta, rice, flour, milk, water, etc.) and nutrients (macronutrients, as carbohydrates, protein, fat, fiber, water; and micronutrients, as minerals and vitamins), meaning that there is a wide range of food types and nutrients an intervention can focus on, escalating the possibilities of interventions in this area and, thus, allowing for a fair level of heterogeneity between them.

Moreover, children are usually not directly responsible for their food buying and consumption. Parents (family context) and the school (educational context) are frequently responsible, for the most part, for children's eating habits. Therefore, to produce stronger positive effects, it may be relevant to widen the scope of interventions to include the community, the family, and multiple environments in which children are involved. Considering individuals exist immersed in social and cultural contexts, changing individual-level behaviors requires paying attention, not only to the individuals but also to the environment where the individual is (Golden & Earp, 2012; McLeroy, Norton, Kegler, Burdine, & Sumaya, 2003; Sallis & Owen, 2015). Nutrition practices are often rooted in broader contexts, such as the family, the community, the society, and even the socioeconomic level and educational level. Acting at different layers and both at the micro- and macro-level allows addressing health problems from a broader perspective, understanding the influence between multiple environments, namely when addressing children, as they are not always responsible for their food choices.

The goal of health interventions, particularly when dealing with children, is not only to affect behaviors at a particular moment but to foster knowledge, understanding, and, ultimately, personal and cultural meaning. This is achieved by developing identities (in this case in regard to nutrition, what one likes more or less, what one identifies with in regard to food consumption, what is healthy for me according to my life goals), contributing to health changes and health behaviors that last a lifespan because they are embedded in the self, in the nature of the person, and in his/her identity. The expression "we are what we eat" could not make more sense in this context, meaning we eat according to our beliefs, our meanings, our identity, our knowledge, and our emotions; in the end, we eat according to who we are.

5 Active learning

5.1 Suggested teaching assignments

1. Search for nutritional health programs with children and identify (1) health outcomes and nutritional objectives; (2) the most common settings; (3) communication strategies and tools that are used to promote communication; (4) behavior change theories and techniques applied; and (5) the stages of health behavior change before and after the intervention.

2. Read about socioecological health interventions. Plan a socioecological school-based food intervention. Consider the various environments children are involved in and include these in the intervention, like school, after-school programs, summer camps, social groups (online and offline), family, institutions, public spaces, community centers, libraries, grocery stores, restaurants/cafeteria/canteen, and communication and media channels. Design a program that involves all these levels/environments in the intervention. Rethink the impact of each of these levels individually and collectively.

3. Practical exercise—develop a practical tool, instrument, and strategy to promote nutritional health behavior change regarding fruit and vegetables with children aged 6 to 10 years old. Focus on a cognitive and behavioral approach. You might search about the food pyramid, the food plate, school meals provision, food supply in schools, and nutritional education to be aware of different tools already in use in nutritional interventions with children. Your practical tool might include any of these instruments or combine two or more, or you may create a new one or a new way of using it/transforming it.

5.2 Recommended readings

Fertman, C. & Allenworth, D. (2010). Health Promotion Programs. From theory to practice. Society of Public Health Education. Jossey-Bass: A Wiley Imprint.

Golden, S. & Earp, J. (2012). Social Ecological Approaches to Individuals and Their Contexts: Twenty Years of Health Education & Behavior Health Promotion Interventions. Health Education & Behavior, 39(3), 364–372. https://doi.org/10.1177/1090198111418634

Henriques, S. & Damasio, M.J. (2021). Re-thinking health behaviour change models from an integrated perspective: building from complementary to reach a unified overview. In C. Vaz de Almeida & S. Ramos (EDS). Handbook of Research on Assertiveness, Clarity and Positivity in Health Literacy (Chapter 6). IGI Global. DOI: https://doi.org/10.4018/978-1-7998-8824-6.

Langford, R., Bonell, C., Jones, H., Pouliou, T., Murphy, S., Waters, E. Campbell, R. (2015). The World Health Organization's Health Promoting Schools framework: a Cochrane systematic review and meta-analysis. BMC public health, 15, 130. doi:https://doi.org/10.1186/s12889-015-1360-y

McLeroy, K. R., Norton, B. L., Kegler, M. C., Burdine, J. N., & Sumaya, C. V. (2003). Community-based interventions. American journal of public health, 93(4), 529–533. doi:https://doi.org/10.2105/ajph.93.4.529

Street, R., Gold, W. & Manning, T. (2013). Health promotion and interactive technology: Theoretical applications and future directions. London, UK: Routledge.

5.3 Case study

Large-scale programs (national level) are often based on policy development and implementation. These programs are mainly initiated and run at local levels on a large scale, to reach national coverage, promoting links at different levels in between the local and national levels. For these programs, both horizontal and vertical organizations and institutions are involved in a coordinated action. Health interventions and particularly healthy nutrition interventions are necessary early in childhood and adolescence to prevent or overturn the unfavorable health effects of overweight and poor eating habits. Schools provide a valued setting to reach large numbers of children, school staff, families, and community members and promote health interventions.

Country A has been promoting a number of nutrition policies and national health initiatives in an attempt to foster a healthy, inclusive, and safe food environment at schools (local level) that can be expanded to families and communities (broader level). These policies involve not only the food sector but also the health, agriculture, marketing, and communication sector, among others.

Here is a list of national initiatives and policies developed and implemented in this context:

- The constitution of the National Authority for Food and Economic Security (2011) that focuses on inspecting food quality, safety, preparation, and conservation in all commercial places that sell or offer food, including schools.
- The development and implementation of the National Program for the Promotion of Healthy Eating (2012) to combat overweight and obesity at the national level, mainly conducted at local health centers.
- The development and implementation of the National Integrated Strategy for the Promotion of Health Eating (2017), focusing on a joint effort between the health and education sector, promoting the presence of health professionals and local health initiatives (such as vaccination, physical activity, healthy eating habits, etc.) in schools.

– Implementation of the regulation on food products available in vending machines at schools and limiting the products available based on food quality and nutritional ingredients for children (2016).
– Commitment to reduce commercial sugar packages (2016), commitment to reduce the salt content of bread (2016).
– Added tax on high-sugar products (2016).
– Mandatory requirement for vegetarian and vegan food options at schools (2017).
– The increase of free distribution of fruit and vegetables in schools (2018).
– Communication policy regulating the restriction on advertising aimed at children under 16 of food and beverages that contain high energy value, salt, sugar, and saturated or processed fat acids in schools, near schools, and on television or cinema in children's programs (2019).

(1) Identify the different sectors involved in these regulations and how they are integrated to reach the same common goal of healthy nutrition.
(2) Plan a socioecological program that considers these national initiatives and proposes how to involve other environments, from local to central organizations and institutions, to foster healthier nutrition habits in the general population, from children to adults and elderly populations.
(3) Identify the different agents and levels that might have an influence on nutrition behavior change and healthy nutrition habits (from the national level to horizontal and vertical organizations, institutions, communities, and social groups, as well as at the individual level). Use those agents to design a social network of relationships between the different agents of social change, starting from the wider level (collective), which should be placed at the center, to the narrower level (the individual), which represents the ramifications and arms of the social network. Like in a sociogram or social network analysis. Here is an image as an example:

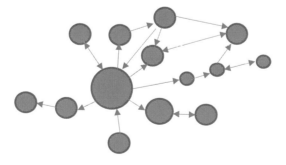

The agents will be the nodes (dots) of your social network and the ties, edges, or links between them represent the connections/influence. You can use arrows in the connection to illustrate the direction of the influence, whether you believe it is one directional or goes both ways.

Use this design when you plan health nutrition intervention in the future, considering all the levels and agents that have an impact on your outcome, and try to consider them in your intervention.

5.4 Titles for research essays

Game-based learning/gamification:
- Costa, C., Tyner, K., Henriques. S., & Sousa, C. (2018). Game Creation in Youth Media and Information Literacy Education. International Journal of Game-Based Learning.
- Henriques, S., Sousa, C. & Costa, C. (2017). Drivers and motivations to game-based learning approaches—a perspective from teachers and parents. INTED2017 proceedings. DOI: 10.21125/inted.2017.1342. Part of ISBN: 9788461784912
- Costa, c., Tyner, K., Henriques, S. & Sousa, C. (2017). Games for media and information literacy—developing media and information literacy skills in children through digital games creation. EDULEARN17. DOI: 10.21125/edulearn.2017.1627. *Part of* ISBN: 9788469737774
- Tobias S., Fletcher J.D., Wind A.P. (2014) Game-Based Learning. In: Spector J., Merrill M., Elen J., Bishop M. (eds) Handbook of Research on Educational Communications and Technology. Springer, New York, NY. https://doi.org/10.1007/978-1-4614-3185-5_38
- Perrotta, C., Featherstone, G., Aston, H. and Houghton, E. (2013). Game-based Learning: Latest Evidence and Future Directions (NFER Research Programme: Innovation in Education). Slough: NFER.
- Huizenga, J., Admiraal, W., Akkerman, S. and Dam, G.t. (2009), Mobile game-based learning in secondary education: engagement, motivation and learning in a mobile city game. Journal of Computer Assisted Learning, 25: 332–344. https://doi.org/10.1111/j.1365-2729.2009.00316.x
- Burke, B. (2014). *Gamify: How gamification motivates people to do extraordinary things.* Bibliomotion, books + media. Routledge - Taylor and Francis Group.
- Plass, J. L., Mayer, R. E., & Homer, B. D. (Eds.). (2020). *Handbook of game-based learning.* The MIT Press.

Health behavior models:
- Bandura, A. (1998). Health promotion from the perspective of social cognitive theory. *Psychology & Health, 13*(4), 623–649. https://doi.org/10.1080/08870449808407422
- Belim, C., & Vaz de Almeida, C. (2018). Healthy thanks to communication: A model of communication competences to optimize health literacy—Assertiveness, clear

language and positivity. In V. E. Papalois & M. Theodosopoulou (Eds.), *Optimizing Health Literacy for Improved Clinical Practices:* IGI Global. https://doi.org/10.4018/978-1-5225-4074-8

- Davis, R., Campbell, R., Hildon, Z., Hobbs, L., & Michie, S. (2015). Theories of behaviour and behaviour change across the social and behavioural sciences: A scoping review. *Health Psychology Review*, *9*(3), 323–344. https://doi.org/10.1080/17437199.2014.941722
- Henriques, S. & Damasio, M.J. (2021). Re-thinking health behaviour change models from an integrated perspective: building from complementarity to reach a unified overview. In C. Vaz de Almeida & S. Ramos (EDS). Handbook of Research on Assertiveness, Clarity and Positivity in Health Literacy (Chapter 6). IGI Global. DOI: https://doi.org/10.4018/978-1-7998-8824-6.
- Michie, S. (2014). *ABC of behaviour change theories: An essential resource for researchers, policy makers and practitioners : 83 theories.*
- Prestwich, A., Sniehotta, F. F., Whittington, C., Dombrowski, S. U., Rogers, L., & Michie, S. (2014). Does theory influence the effectiveness of health behavior interventions? Meta-analysis. *Health Psychology*, *33*(5), 465–474. https://doi.org/10.1037/a0032853
- Sallis, J. F., & Owen, N. (2015). Ecological models of health behavior. In *K. Glanz, B. K. Rimer, & K. 'V.' Viswanath (Eds.), Health Behavior and Health Education: Theory, Research and Practice* (pp. 43–64). Jossey-Bass/Wiley.
- Schwarzer, R. (2016). Health Action Process Approach (HAPA) as a Theoretical Framework to Understand Behavior Change. *Actualidades En Psicología*, *30*(121), 119. https://doi.org/10.15517/ap.v30i121.23458

5.5 Recommended projects URL

GamiLearning—https://cicant.ulusofona.pt/research/projects/european-funding/288-gamilearning.

Improving Prenatal Health Communication: Engaging Men via e-Health https://cicant.ulusofona.pt/research/projects/european-funding/359-improving-prenatal-health-communication-engaging-men-via-e-health.

Delivering Repeated Health Messages https://cicant.ulusofona.pt/research/projects/european-funding/358-delivering-repeated-health-messages-through-digital-media-to-increase-physical-activity-in-dialysis-patients.

Annex 1

Citation	Purpose of the study	Research design	Participants (age, gender, country)	Intervention	Theoretical models	Measures	Outcomes	Media used
Banos et al. (2013)	To study the efficacy and acceptability of an online game called ETIOBE Mates, designed to improve children's nutritional knowledge	Quasi-experimental	N (intervention): 73 N (control): 155 School setting: 4th to 6th grade Age: 10–13; (mean = 11.2) Sex: male and female Location: Valencia, Spain	**Intervention:** videogame ETIOBE Mates Treatment: play ETIOBE Mates Control: traditional paper and pencil mode of information (pamphlet called "balanced diet") **Type:** diet **Duration:** 2 weeks **Topics:** nutritional knowledge— nutritional terms, awareness of dietary recommendations, nutrients contained in food, practical food choices, awareness of diet–disease associations.	Not reported	• BMI • Internet and game-playing habits (6-item questionnaire) • Nutritional Knowledge Questionnaire for children (35 items) • Acceptability-playability questionnaire: 26 items	Anthropometry (height, weight BMI) Internet and game-playing habits Nutritional Knowledge	Online videogame
Baranowski et al. (2011)	To evaluate the outcome of playing two video games on children's diet, physical exercise, and adiposity.	RCT	N (intervention): 103 N (control): 50 School setting: middle school Age: 10–12 Sex: male and female Location: North Carolina and Texas, United States	**Intervention:** Videogames—*Escape from Diab* and *Nanoswarm; Invasion from inner space* Treatment: played both videogames in sequence during 9 sessions of ±40 min each (a total of ~6 h of play) Control: played diet and PA knowledge-based games. **Type:** Diet + PA **Topics:** diet, FV, adiposity, physical activity	Serious videogames Game-based learning	• Three noncon-secutive days of 24 h dietary recalls • Five consecutive days of PA using accelerometers • Height and weight • Waist circumfer-ence, triceps skinfold.	Serving of FV and water Minutes of moderate PA	Online videogames

Study	Aim	Design	Sample	Intervention	Theory	Measurement	Outcome	Delivery
Bech-Larsen and Gronhoj (2013)	Feedback intervention aimed at increasing consumption of FV.	RCT	N (intervention): 169 N (control): 87 School setting: middle school Age:12 Sex: male and female Location: Denmark	**Intervention:** mobile phone-based SMS system Treatment: nutrition education lessons with a dietician and 4 weeks of SMS diary Control: only nutrition classes with a dietician **Type:** diet **Duration:** 4 weeks **Topics:** FV, goals, planning, how to count units/portions	Social cognitive theory Goal theory (goal setting strategy)	• FV consumption	FV	SMS
Cullen et al. (2016)	To increase FV consumption.	RCT	N (intervention): 290 N (control): 97 School-setting: 4th and 5th grade Age (mean): 9–11 Sex: male and female Location: Houston, United States	**Intervention:** "Squire's Quest! 11: saving the Kingdom of Fivealot!" Four experimental groups based on the type of implementation intentions used: (1) control (none); (2) action; (3) coping; (4) action + coping. Treatment: play the game and implementation intentions (action, coping, action + coping) Control: none **Type:** diet **Duration:** 6 months **Topics:** setting meal-specific goals for eating FV (breakfast, lunch, snacks, and dinner). Setting a personal plan to eat 5 FV daily.	Implementation intention, action plans, coping plans.	• Children completed three 24-h dietary recalls (2 weekdays, 1 weekend day) via phone (breakfast, lunch, snack, dinner intakes). • FV intake in those 3 days was averaged to estimate the amount of VF consumption.	FV consumption	Online videogame

Continued

Citation	Purpose of the study	Research design	Participants (age, gender, country)	Intervention	Theoretical models	Measures	Outcomes	Media used
Ezendam et al. (2012)	To evaluate the short and long-term results of FATaintPHAT, a web-based computer-tailored intervention aiming to increase physical activity and healthy eating.	Cluster Randomized Controlled Trial	N (intervention): 485 N (control): 398 School setting: secondary education schools Age: 12–13 Sex: male and female Location: Netherlands	**Intervention:** FATaintPHAT—a web-based computer-tailored intervention. Treatment: a web-based computer-tailored intervention Control: regular curriculum **Type:** diet + PA **Duration:** 8 lessons online (Internet) over 10 weeks/2-year follow-up **Topics:** weight management and energy balance-related behaviors.	Theory of Planned behavior Precaution Adoption Process model Implementation intentions	• Self-reported behaviors • Food frequency questionnaire • 24 h recall for snacks and FV • Flemish validated questionnaire for PA — Pedometers • Number of days with moderate to vigorous PA—Pedometers • Anthropometric measures using digital scales	Self-reported behaviors for diet, PA, and sedentary behavior; Pedometer count. Anthropometry (BMI, waist circumference)	Computer/ Internet
Fassnacht et al. (2015)	To examine the efficacy, satisfaction, and adherence of a mobile phone SMS service to promote health behaviors.	RCT	N (intervention): 22 N (control): 27 School-setting: 4th grade Age: 8–10 (mean = 9.6; SD = 0.4) Sex: male and female Location: Portugal	**Intervention:** SMS-based monitoring and feedback system Treatment: SMS intervention Control: No SMS **Type:** Diet + PA **Duration:** 8 weeks **Topics:** diet, physical activity, screen time	Social cognitive theory and behavioral models	• Food frequency questionnaire (FV) • PA and screen time—family eating and activity questionnaire • Anthropometric measures (height, weight, BMI) • Pedometer • Satisfaction questionnaire	FV PA Screen time	SMS

| Jones et al. (2014) | To explore a low-cost, game-based intervention to increase FV consumption. | Alternating-treatments design trial | N (intervention): 251
N (control): 251
School setting: 5th grade
Age: 10–11
Sex: male and female
Location: Utah, United States | Intervention: game-based intervention—FIT Game. The whole school played the game. At baseline, 15 days to measure FV consumption using FV waste (weight-based measure). FV were targeted each day for increasing consumption. A story with heroes and villains was told to students and they were asked to help by eating more of the target food during lunch. Reinforcement was done when students met the goal.
Type: diet
Duration: 1.5 months (44 days)
Topics: story with heroes and villains, FV, healthy eating. | Incentives model to increase target behavior; gamification, competition/collaboration | • FV weighted (plate–waste measure technique) | FV intake | Game (collaborative, competitive game) |

Continued

Citation	Purpose of the study	Research design	Participants (age, gender, country)	Intervention	Theoretical models	Measures	Outcomes	Media used
Joyner et al. (2017)	To analyze whether FIT Game can be administered and keep its efficacy without teachers' work, presenting all materials on visual displays at the school cafeteria.	Alternating-treatments design trial	N (intervention): 572 N (control): 572 School setting: 5th grade Age: 6–11 Sex: male and female Location: Utah, United States	**Intervention:** game-based intervention—FIT Game. The whole school played the game. At baseline, 15 days to measure FV consumption using FV waste (weight-based measure). FV were target each day for increasing consumption. A story with heroes and villains was shown to students on screens at cafeteria to students and they were asked to help by eating more of the target food during lunch. Reinforcement was done when students met the goal. No teacher intervention. **Type:** diet **Duration:** 1.5 months (44 days) **Topics:** story with heroes and villains, FV, healthy eating.	Incentives model to increase target behavior; gamification, competition/collaboration	• FV weighted (plate-waste measure technique)	FV intake	Game (collaborative, competitive game)

Study	Aim	Design	Sample/Setting	Intervention	Theory	Measures	Outcomes	Technology
Moore et al. (2009)	To improve children's nutrition-related self-care operations.	Quasiexperimental	N (intervention): 64 N (control): 62 School setting: 5th grade Age: 9–11 Sex: male $n=46$, Female $n=80$ Location: Washington DC, United States	**Intervention:** COLOR MY PYRAMID Treatment: education and activity content + experiential learning with digital game MyPyramid Blast Off Game Control: education and activity content + experiential learning using individual computers to evaluate their diets in small groups **Type:** diet + PA **Duration:** 6 classes—3 months **Topics:** nutrition concepts, moderation, variety, portion sizes, exercise.	Self-care deficit nursing theory (Orem, 2001)	• Nutrition knowledge questionnaire (28 items) • Adolescent nutrition self-care questionnaire (50 items, Likert scale) (Moore et al., 2009) • Student Physical activity—Recall of minutes of physical activity per day • Nutrition Status—research anthropometric Form (RA) (7 items)—record of height, weight, blood pressure	• Nutrition knowledge Nutrition self-care practices PA, Anthropometry (blood pressure, height, weight, BMI)	Online videogame
Nollen et al. (2014)	To test the feasibility and efficacy of mobile technology interventions to promote FV consumption and decrease sweetened beverages.	RCT	N (intervention): 26 N (control): 25 School setting: after-school program Age (mean): 9–14 Sex: female only Location: Kansas, United States	**Intervention:** mobile technology Treatment: mobile technology use with real-time setting prompts, self-monitoring, tips, feedback, positive reinforcement related to target behaviors Control: written manual with same content, no prompting	Behavioral weight control principles	• Two-day 24h dietary recall (1 weekday and weekend day) • Screen time was assessed using Brief Questionnaire of Television Viewing and Computer Use • Weight and height	Device use Effect sizes of FV, sugar-sweetened beverages, screen time, BMI	Mobile app with prompting SMS

Continued

Citation	Purpose of the study	Research design	Participants (age, gender, country)	Intervention	Theoretical models	Measures	Outcomes	Media used
Pedersen et al. (2016)	To examine the effects of a feedback intervention employing text messaging on adolescents' behavior, self-efficacy, and outcome expectations regarding FV	RCT	N (intervention): 489 + 497 N (control): 502 School setting: Secondary school Age: 11–16 (mean = 12.9) Sex: male and female Location: Denmark	**Intervention:** feedback intervention with text messaging Treatment: (1) text messaging (2) text messaging + educational group class. Control: no text messaging **Type:** diet **Duration:** 11 weeks (weeks 1 and 11 to pre- and posttests) **Topics:** fruit and vegetable intake	Self-efficacy— Social Cognitive Theory	• Participants' behavior: Survey with intake frequency items and FV • Self-efficacy: 9 items adapted from Perry et al.'s (2008) Physical Activity and Healthy Food Efficacy Scale for Children (PAHFE). • Outcome expectations: instrument adapted from Thøgersen and Grønhøj (2010). Engagement: measured by participants' total SMS sent.	FV intake Self-efficacy Outcome expectations	SMS

Additional intervention details (upper portion): messages **Type:** diet **Duration:** 12 weeks **Topics:** FV, sugar-sweetened beverages, screen time

Study	Objective	Design	Sample/Setting	Intervention	Theory	Measures	Outcomes	Mode
Rees et al. (2010)	To evaluate the effectiveness of a computer-generated tailored intervention leaflet aimed at increasing brown bread, wholegrain cereal, FV	Clustered RCT	N (intervention): 323 N (control): 314 School setting: secondary schools Age: 12–16 (mean age=14) Sex: female only Location: London and West Midlands, United Kingdom	**Intervention:** computer-generated tailored intervention leaflet Treatment: tailored leaflet made by using a computer program that automatically considered and analyzed questionnaire responses and produced a leaflet using a standard template. Control: traditional leaflet using national guidelines of intake recommendations. **Type:** diet **Duration:** 4.5 months (3 months intervention +3 weeks for baseline +3 weeks for endline dietary recalls) **Topics:** fruit and vegetable intake, portions, planning, and implementation.	Theory of Planned Behavior	• Dietary intake was accessed via three 24h dietary recalls (previous day/3 different days at baseline and then a follow-up again). • Psychological questionnaire (beliefs) (30 items for each food)	Brown bread, wholegrain cereals, fruit, vegetables. Psychological outcomes Frequency measure of target food consumption	Computer-generated leaflet tailored
Silva et al. (2015)	To evaluate the effectiveness of the program to increase FV consumption and physical activity.	RCT	N (intervention): 69 N (control): 70 School setting: elementary school Age: 8–10 (mean age = 9.9) Sex: male and female Location: Braga, Portugal	**Intervention:** Text messaging, SMS Intervention: 2 × 60-min psycho-educational sessions +SMS program (to monitor their behavior for 8 weeks providing	Social cognitive theory and behavioral models	• Food frequency and physical activity questionnaire • Food Frequency Questionnaire (Vereecken & Maes, 2003)	BMI FV intake Exercise Screen time	SMS

Continued

Citation	Purpose of the study	Research design	Participants (age, gender, country)	Intervention	Theoretical models	Measures	Outcomes	Media used
				information at the end of the day; by SMS: how many FV they have eaten; how many steps they have walked; how many minutes they have spent in front of a screen) Control: only 2 × 60-min psycho-educational sessions Type: diet + PA Duration: 8 weeks + follow-up (4 weeks later) Topics: FV, screen time, physical activity		• Family Eating and Activity Habits Questionnaire (Golan & Weizman, 1998) • Pedometers (to count steps per day) • BMI (height and weight) • Self-report satisfaction questionnaire (after the intervention for intervention group)		
Sharma et al. (2015)	Evaluate the effects of the Quest to Lava Mountain (QTLM) computer game on dietary behaviors and physical activity.	Quasiexperimental	N (intervention): 44 N (control): 50 School setting: 4th and 5th grade Age:9–10 Sex: male and female Location: Texas, United States	Intervention: Quest to Lava Mountain— game in which players must create an avatar and make it eat healthy and stay active. Treatment: play the game Control: don't play the game Type: diet + PA Duration: recommended exposure duration was 90min/week for 6 weeks. Topics: food, relationship between	Social cognitive theory, Theory of Reasoned Action	• Two random 24-h recalls and self-report surveys	Diet (FV, sugar, fiber, fat, carbohydrates, protein, calcium) PA Psychosocial factors	Computer-based educational games

Continued

| Struempler et al. (2014) | To increase FV consumption in youth | Quasiexperimental | N (intervention): 1674
N (control): 803
School setting: 3rd grade students
Age: 8 to 9
Sex: 51% male, 49% female
Location: Alabama, United States | food and exercise, healthy food and beverages, whole grains, low-fat dairy, water, healthy diet based on variety and moderation, physically active lifestyle helps maintain health
Intervention: BODY QUEST Food of the warrior Experimental: 6 traditional educator-led lessons + 1 week of nontraditional lesson with iPad apps + FV tasting + take-home activities
Control: no intervention.
Type: diet
Duration: 17 weeks/45-min classes
Topics: trying new foods, food groups, balanced meals, food nutrients, healthy snacks, extending FV message to others. | Experiential Learning Theory | • Checklist developed for the study: "What's for lunch (W4L)?" checklist
• Students self-report | Weekly FV consumed in school lunch | iPad apps |

Citation	Purpose of the study	Research design	Participants (age, gender, country)	Intervention	Theoretical models	Measures	Outcomes	Media used
Yang et al. (2017)	To identify the effectiveness of an obesity prevention program focused on motivating environments	Quasiexperimental	N (intervention): 418 N (control): 350 Education level: 4th and 7th grade Age: 9 to 10 and 12 to 13 Sex: male and female Location: Chungju, Korea	**Intervention:** Environmental intervention. Treatment: diet and exercise-related educational video content provided by Internet protocol TV (IPTV) during rest time + design materials/publicity exposed in school. Control: usual school curriculum/ traditional learning **Type:** diet + PA **Duration:** 1 year **Topics:** educational dietary, daily lunch menu, exercise.	Not reported	• Anthropometry measures • Student Physical Activity Questionnaire about physical activity, sleeping, dietary, nutritional knowledge, self-image, self-respect, life quality, depression. • Parental questionnaire (9 items about the child and 11 items about the child's family)	Anthropometry (height, weight, BMI Waist size, blood pressure) PA Nutritional behavior	IPTV—TV video content; printed posters

Acknowledgments

Work founded by the Science and Technology Foundation (Fundação para a Ciência e a Tecnologia—Portugal), under the grant SFRH/BD/13201172017.

References

Bandini, L., Curtin, C., Hamad, C., Tybor, D., & Must, A. (2005). Prevalence of overweight in children with developmental disorders in the continuous national health and nutrition examination survey (NHANES). *The Journal of Pediatrics, 146,* 738–743.

Banos, R. M., Cebolla, A., Oliver, E., Alcaniz, M., & Botella, C. (2013). Efficacy and acceptability of an Internet platform to improve the learning of nutritional knowledge in children: The ETIOBE mates. *Health Education Research, 28*(2), 234–248. https://doi.org/10.1093/her/cys044.

Baranowski, T., Baranowski, J., Thompson, D., Buday, R., Jago, R., Griffith, M. J., … Watson, K. B. (2011). Video game play, child diet, and physical activity behavior change. *American Journal of Preventive Medicine, 40*(1), 33–38. https://doi.org/10.1016/j.amepre.2010.09.029.

Bech-Larsen, T., & Grønhøj, A. (2013). Promoting healthy eating to children: A text message (SMS) feedback approach: Promoting healthy eating to children. *International Journal of Consumer Studies, 37*(3), 250–256. https://doi.org/10.1111/j.1470-6431.2012.01133.x.

Brug, J., Campbell, M., & van Assema, P. (1999). The application and impact of computer-generated personalized nutrition education: A review of the literature. *Patient Education and Counseling, 36*(2), 145–156. https://doi.org/10.1016/s0738-3991(98)00131-1.

Cane, J., O'Connor, D., & Michie, S. (2012). Validation of the theoretical domains framework for use in behaviour change and implementation research. *Implementation Science, 7*(1), 37. https://doi.org/10.1186/1748-5908-7-37.

Cullen, K. W., Liu, Y., & Thompson, D. I. (2016). Meal-specific dietary changes from squires Quest! II: A serious video game intervention. *Journal of Nutrition Education and Behavior, 48*(5), 326–330.e1. https://doi.org/10.1016/j.jneb.2016.02.004.

De Bourdeaudhuij, I., Van Cauwenberghe, E., Spittaels, H., Oppert, J., Rostami, C., Brug, J., … Maes, L. (2010). School-based interventions promoting both physical activity and healthy eating in Europe: A systematic review within the HOPE project. *Obesity Reviews, 12,* 205–216. https://doi.org/10.1111/j.1467-789X.2009.00711.x.

Ezendam, N. P. M., Brug, J., & Oenema, A. (2012). Evaluation of the web-based computer-tailored FATaintPHAT intervention to promote energy balance among adolescents: Results from a school cluster randomized trial. *Archives of Pediatrics and Adolescent Medicine, 166*(3), 248–255. https://doi.org/10.1001/archpediatrics.2011.204.

Fassnacht, D. B., Ali, K., Silva, C., Gonçalves, S., & Machado, P. P. P. (2015). Use of text messaging services to promote health behaviors in children. *Journal of Nutrition Education and Behaviour, 47*(1), 75–80. https://doi.org/10.1016/j.jneb.2014.08.006.

Fertman, C., & Allenworth, D. (2010). *Health promotion programs. From theory to practice.* Society of Public Health Education. Jossey-Bass: A Wiley Imprint.

Golan, M., & Weizman, A. (1998). Reliability and validity of the Family Eating and Activity Habits Questionnaire. *European Journal of Clinical Nutrition, 52*(10), 771–777. https://doi.org/10.1038/sj.ejcn.1600427.

Golden, S., & Earp, J. (2012). Social ecological approaches to individuals and their contexts: Twenty years of health education & behavior health promotion interventions. *Health Education & Behavior, 39*(3), 364–372. https://doi.org/10.1177/1090198111418634.

Henriques, S., & Damasio, M. J. (2021). Re-thinking health behaviour change models from an integrated perspective: Building from complementarity to reach a unified overview. In C. Vaz de Almeida, & S. Ramos (Eds.), *Handbook of research on assertiveness, clarity and positivity in health literacy* IGI Global. https://doi.org/10.4018/978-1-7998-8824-6 (Chapter 6).

Higgins, J. P. T., López-López, J. A., Becker, B. J., Davies, S. R., Dawson, S., Grimshaw, J. M., … Caldwell, D. M. (2019). Synthesising quantitative evidence in systematic reviews of complex health interventions. *BMJ Global Health*, *4*(Suppl. 1), e000858. https://doi.org/10.1136/bmjgh-2018-000858.

Jones, B. A., Madden, G. J., & Wengreen, H. J. (2014). The FIT game: Preliminary evaluation of a gamification approach to increasing fruit and vegetable consumption in school. *Preventive Medicine*, *68*, 76–79. https://doi.org/10.1016/j.ypmed.2014.04.015.

Joyner, D., Wengreen, H. J., Aguilar, S. S., Spruance, L. A., Morrill, B. A., & Madden, G. J. (2017). The FIT game III: Reducing the operating expenses of a game-based approach to increasing healthy eating in elementary schools. *Games for Health Journal*, *6*(2), 111–118. https://doi.org/10.1089/g4h.2016.0096.

Langford, R., Bonell, C., Jones, H., Pouliou, T., Murphy, S., Waters, E., & Campbell, R. (2015). The World Health Organization's health promoting schools framework: A Cochrane systematic review and meta-analysis. *BMC Public Health*, *15*, 130. https://doi.org/10.1186/s12889-015-1360-y.

McLeroy, K. R., Norton, B. L., Kegler, M. C., Burdine, J. N., & Sumaya, C. V. (2003). Community-based interventions. *American Journal of Public Health*, *93*(4), 529–533. https://doi.org/10.2105/ajph.93.4.529.

Monteiro, R. (2010). *Metodologias de meta-análise aplicadas nas ciências da saúde. Dissertação de Mestrado em aplicações estatísticas às ciências da saúde, da vida e do ambiente.* https://ubibliorum.ubi.pt/bitstream/10400.6/1849/1/Meta-An%C3%A1lise.pdf. (Accessed 13 March 2022.)

Moore, J. B., Pawloski, L. R., Goldberg, P., Kyeung, M. O., Stoehr, A., & Baghi, H. (2009). Childhood obesity study: A pilot study of the effect of the nutrition education program color my pyramid. *The Journal of School Nursing*, *25*(3), 230–239. https://doi.org/10.1177/1059840509333325.

Nollen, N. L., Mayo, M. S., Carlson, S. E., Rapoff, M. A., Goggin, K. J., & Ellerbeck, E. F. (2014). Mobile technology for obesity prevention. *American Journal of Preventive Medicine*, *46*(4), 404–408. https://doi.org/10.1016/j.amepre.2013.12.011.

Orem, D. E. (2001). *Nursing concepts of practice* (6th). St. Louis, MO: Mosby.

Oude Luttikhuis, H., Baur, L., Jansen, H., Shrewsbury, V., O'Malley, C., Stolk, R., & Summerbell, C. (2009). Interventions for treating obesity in children. *Cochrane Database of Systematic Reviews*, (1), CD001872. https://doi.org/10.1002/14651858.CD001872.pub2.

Painter, J. E., Borba, C. P. C., Hynes, M., Mays, D., & Glanz, K. (2008). The use of theory in health behavior research from 2000 to 2005: A systematic review. *Annals of Behavioral Medicine*, *35*(3), 358–362. https://doi.org/10.1007/s12160-008-9042-y.

Pedersen, S., Grønhøj, A., & Thøgersen, J. (2016). Texting your way to healthier eating? Effects of participating in a feedback intervention using text messaging on adolescents' fruit and vegetable intake. *Health Education Research*, *31*(2), 171–184. https://doi.org/10.1093/her/cyv104.

Perry, J., Dewine, D., Duffy, R., & Vance, K. (2008). The Academic self-efficacy of urban youth. A mixed-methods study of a school-to-work program. *Journal of Career Development*, *34*(2), 103–126. https://doi.org/10.1177/0894845307307470.

Rees, G., Bakhshi, S., Surujlal-Harry, A., Stasinopoulos, M., & Baker, A. (2010). A computerised tailored intervention for increasing intakes of fruit, vegetables, brown bread and wholegrain cereals in adolescent girls. *Public Health Nutrition*, *13*(8), 1271–1278. https://doi.org/10.1017/S1368980009992953.

Richardson, M., Garner, P., & Donegan, S. (2019). Interpretation of subgroup analyses in systematic reviews: A tutorial. *Clinical Epidemiology and Global Health*, *7*(2), 192–198. https://doi.org/10.1016/j.cegh.2018.05.005.

Sahoo, K., Sahoo, B., Choudhury, A. K., Sofi, N. Y., Kumar, R., & Bhadoria, A. S. (2015). Childhood obesity: Causes and consequences. *Journal of Family Medicine and Primary Care*, *4*(2), 187–192. https://doi.org/10.4103/2249-4863.154628.

Sallis, & Owen. (2015). Ecological models of health behavior. In Galnz, Rimer, & Viswanath (Eds.), *Health behavior—Theory, research and practice* (5th ed.). Jossey-Bass—A Wiley brand.

Sharma, S. V., Shegog, R., Chow, J., Finley, C., Pomeroy, M., Smith, C., & Hoelscher, D. M. (2015). Effects of the Quest to Lava Mountain computer game on dietary and physical activity behaviors of elementary school children: A pilot group-randomized controlled trial. *Journal of the Academy of Nutrition and Dietetics*, *115*(8), 1260–1271. https://doi.org/10.1016/j.jand.2015.02.022.

Silva, C., Fassnacht, D. B., Ali, K., Gonçalves, S., Conceição, E., Vaz, A., … Machado, P. P. (2015). Promoting health behaviour in Portuguese children via short message service: The efficacy of a text-messaging programme. *Journal of Health Psychology*, *20*(6), 806–815. https://doi.org/10.1177/1359105315577301.

Street, R., Gold, W., & Manning, T. (2013). *Health promotion and interactive technology: Theoretical applications and future directions*. London, UK: Routledge.

Struempler, B. J., Parmer, S. M., Mastropietro, L. M., Arsiwalla, D., & Bubb, R. R. (2014). Changes in fruit and vegetable consumption of third-grade students in body Quest: Food of the warrior, a 17-class childhood obesity prevention program. *Journal of Nutrition Education and Behavior*, *46*(4), 286–292. https://doi.org/10.1016/j.jneb.2014.03.001.

Thogersen, J., & Gronhoj, A. (2010). Eletricity saving in households – A social cognitive approach. *Emergy Policy*, *38*(12), 7732–7743. https://doi.org/10.1016/j.enpol.2010.08.025.

Vereecken, C. A., & Maes, L. (2003). A Belgian study on the reliability and relative validity of the Health Behaviour in School-Aged Children food-frequency questionnaire. *Public Health Nutrition*, *6*(6), 581–588. https://doi.org/10.1079/phn2003466.

Whittemore, R., Jeon, S., & Grey, M. (2013). An internet obesity prevention program for adolescents. *Journal of Adolescent Health*, *52*(4), 439–447. https://doi.org/10.1016/j.jadohealth.2012.07.014.

World Health Organization. (2022). *Non-communicable diseases. Fact-sheets.*, 1. https://www.who.int/news-room/fact-sheets/detail/noncommunicable-diseases. (Accessed 11 December 2022).

World Health Organization. (2009). *Interventions on diet and physical activity: What works: Summary report*. https://apps.who.int/iris/handle/10665/44140. (Accessed 21 October 2022).

World Health Organization. (2021). *Obesity and overweight. WHO fact-sheets*. https://www.who.int/news-room/fact-sheets/detail/obesity-and-overweight. (Accessed 23 October 2021).

Yang, Y., Kang, B., Lee, E. Y., Yang, H. K., Kim, H.-S., Lim, S.-Y., & Yoon, K.-H. (2017). Effect of an obesity prevention program focused on motivating environments in childhood: A school-based prospective study. *International Journal of Obesity*, *41*(7), 1027–1034. https://doi.org/10.1038/ijo.2017.47.

CHAPTER 10

Positive psychology's role in the training of health professionals: Looking into the future

Helena Águeda Marujo

Instituto Superior de Ciências Sociais e Politicas—ISCSP-CAPP, Center for Administration and Public Policies—CAPP, UNESCO Chair on Education for Global Peace Sustainability, University of Lisbon, Lisbon, Portugal

1 The future of education: Global perspectives

Education is being reshaped. New questions have surfaced in the last decade, namely "What knowledge, skills, attitudes, and values will today's students need to thrive and shape their world?" and "How can instructional systems develop these knowledge, skills, attitudes, and values effectively?" (OECD, 2018).

The need to rethink the contents, the methods, and the stakeholders—the *what*, the *why*, the *how*, the *for whom*, and the *with whom* of education—is paired with discussions on the ontology behind what it means, currently, to train and prepare a person for the labor market, as well as for living a healthy life with quality and worth (UNESCO, 2021). Besides, debating current education means introducing new approaches to prepare for responding effectively to the present and future challenges of our contemporary society (United Nations Children's Fund, 2021).

Knowledge and learning are humanity's highest renewable capitals for answering challenges and discovering new solutions for a harmonious world. Therefore, today's education needs to do more than reply to a fluctuating and unpredictable world. Education must be groundbreaking in its vision and impact, namely anticipating what will be the prerequisites for a healthy, just, safe, ecologically sustainable, inclusive, ethical, happy, and peaceful society (Marujo & Casais, 2021).

These topics are at the center of today's debates on education, long acknowledged as a powerful force for social transformation (UNESCO, 2021). That dimension of social change is also the reason why education is the key piece of the global commitments and expected achievements proposed by the United Nations Organization to shape humanity's horizons (UN General Assembly, 2015). Naturally, there are multiple dimensions to the future and there will likely be various desirable and undesirable futures, but the world is demanding innovative and forward-thinking approaches to find new answers on how knowledge, education, and learning need to be reimagined to address increasing

Active Learning for Digital Transformation in Healthcare Education, Training and Research
https://doi.org/10.1016/B978-0-443-15248-1.00011-4

complexity, uncertainty, and precarity. Learners must not only improve what they know but also develop skills, attitudes, and values that will support them to be proficient people.

Consequently, diverse global initiatives are emerging from around the planet to reimage how knowledge and learning can shape the future of humanity and the planet (Marujo & Casais, 2021). Among the changes proposed, the new visions of the future of learning defend social assets, such as enablers of education equity and quality to ensure excellence in education for everyone, which requires that all individuals have access to the same level of resources, knowledge, and support. This includes not only technology and connectivity but also other relevant assets, like gender parity and excellence in teacher training (UNESCO, 2021). Currently, genuine transformation in education also requires a systems-based approach and rigorous evidence of what works, involving all the stakeholders in collaborative actions while using empirical data to inform contents and pedagogical practices are considered important aims to attain outstanding educational outcomes (OECD, 2018).

Contemporary worth-valued education promotes knowledge and resourcefulness and assures the acquisition of foundational skills (like literacy, numeracy, analytical problem solving and assorted high-level, abstract and reflexive rational skills) (Saliceti, 2015). Nevertheless, that is considered insufficient for the world we live in. There is also a requirement for an education model that matures essential interpersonal, emotional, and social skills and develops the abilities, values, and attitudes that qualify citizens to acknowledge and experience healthy, peaceful, and fulfilled lives for themselves and for all (Marujo, 2021).

Consequently, the role of psychological dimensions in the future of education has gained traction. The need to train for resilience and to promote proficient and sustainable mental health—acknowledging its clear and proven relationships with physical health and the flourishing of human beings and society (Seligman, Ernst, Gillham, Reivich, & Linkins, 2009)—is an aspiration in contemporary educational paradigms (UNESCO, 2021). These foresights are also present in today's health education models.

2 Tackling healthcare education

The coming-up of healthcare education is also at stake (e.g., Faerber et al., 2019; Gebbie, Rosenstock, & Hernandez, 2003; Lucey, Thibault, & Ten Cate, 2018).

Undeniably, today's healthcare industry faces convoluted challenges, alongside unpredictable societal trials, which result in demands for novel models of education that will rethink and redesign the training of healthcare professionals. The aim is that these professionals deliver healthcare in a more effective manner and better tackle the requests of an activity that is expected to revolutionize treatment models and lifestyle behaviors, develop new and innovative diagnostic tools, advance the holistic health of people, and ameliorate healthcare professional-patient relationships.

Alongside the premises of the need for the inclusion of digital methods and blended learning in the training of healthcare practitioners (Liu et al., 2016), today's intricate problems "require multi-faceted public health actions based on an ecological approach to problem solving, which requires a well-educated interdisciplinary cadre of public health professionals who focus on population health and understand the multiple determinants that affect health. A cadre of professionals who also understand that successful interventions require understanding not only of the effects of biology and behavior, but also the social, environmental, and economic contexts within which populations exist." (Gebbie et al., 2003, p. 1).

The psychological preparedness and resilience of the future healthcare practitioner is also at stake, namely what concerns his or her capacity to deal with very stressful and unpredictable health contexts, as the novel SARS-CoV-2 virus pandemic has shown in the last 2 years (e.g., Alsoufi et al., 2020).

The field of health sciences has long been predominantly dedicated to prevention, diagnosis, treatment, and cure of disease. However, health is more than the mere absence of disease. As frequently underscored—but not necessarily applied in the everyday of health practices, nor in the training of health professionals—in, 1946 the World Health Organization (WHO) defined health as "a state of complete physical, mental and social well-being and not merely the absence of disease or infirmity" (World Health Organization, 1946). This definition overlaps the experience of health with that of "well-being," introducing both the social and the psychological dimensions, alongside the somatic one. In the Ottawa Charter for Health Promotion, held in 1986, the same WHO proposed the inclusion of cultural factors for improved well-being. Currently, more than ever, a more amplified concept of health is therefore being debated (Neves, 2021).

Generally speaking, health education has accumulated a tradition characterized by a specific and somewhat rigid format: focused on content and on a pedagogy of transmission, with disconnection between thematic nuclei; excessive workload for certain content and with low or no offer of optional, interdisciplinary topics; disentanglement between teaching, research, and extension; and predominantly using an encyclopedic design and an orientation toward illness and rehabilitation (IOM, 2011). Within the scope of educational policies, the training in the health area has not had an integrative orientation potentializing competencies for integrality and innovation, which includes: the confrontation of the population's health needs; the psychological, social, physical, and cultural necessities of the healthcare practitioners and the population; and the development of a critical stance toward the current health education models and of the existing health systems (Freudenberg, 1978; Jourdan, Samdal, Diagne, & Carvalho, 2008).

In the modern university, epistemological cartography has been constituted into disciplines and departments. This historical cutout was taken as if it was an epistemology, giving rise to the corporativism of specialties and bureaucratic controls, which hinder interdisciplinary practices, accepting fragmentation as a way of organizing knowledge and actions (Carvalho & Ceccim, 2012). Recently, the Liaison Committee on Medical Education, the accrediting body for US and Canadian Medical Schools, recommended that medical education curricula include behavioral and socioeconomic curricula. Yet, there is still a long way toward interdisciplinary training; in particular, many health schools do not focus on a comprehensive approach to health and well-being (Lianov et al., 2020).

Change that includes aggregate well-being as a priority subject is needed to help health students prepare for an evolving healthcare environment, respond to the current mental health crisis, and also comply with increasing student demand for such training, which includes content on behavior change, lifestyle diseases, and emotional well-being (Parkes, Poland, Allison, et al., 2020). Nonetheless, well-being and positive behavioral change are often seen as primary domains of mental health settings, and primary care and other nonmental health care sceneries often miss the domain of subjective well-being and lifestyle behavior and attitudes. This is explained by time limitations, rigidity and conservatism on the contents taught, inadequate incentives, overspecialization, and general lack of training (Lianov et al., 2020).

As a consequence, a global scientific discipline of health—beyond the mere absence of disease—barely exists, although a focus on health rather than illness will be cost-saving and lifesaving (Seligman, 2008). This is the proposal of the evolving notion of positive health that takes a groundbreaking multi- and interdisciplinary approach to health and emotional well-being and centers on upholding people's affirmative health assets—forces that can subsidize for a healthier and longer life.

The science of positive psychology has dedicated itself to this aim in the last 25 years. This outlook will drive an important change in the advancement of healthcare education.

3 Positive psychology: A science of individual and relational strengths

Historically, health disciplines have focused almost exclusively on pathology and rarely on well-being, prevention, and promotion of what we aim to attain (Seligman, 2002; Seligman & Csikszentmihalyi, 2000). In health sciences in general, this influence of the negative can be recognized in view of the amount of research that emphasizes deleterious experiences (for instance, anxiety, burnout or depression in mental sciences, and osteoporosis, ulcers, cancer, and so many associated topics in the physical realm), reinforcing problems and difficulties rather than optimization (Jackson, 2000).

However, at the beginning of the new millennium, a new area of psychology emerged, denominated positive psychology, that proposed a change in focus (Costa & López, 2008). From the point of view of this emerging area, the emphasis should no longer be on studying what was negative (e.g., mental illnesses, difficulties, malaise, all kinds of individual and collective pathologies, and avoidance of them) but should instead become the scientific attention on the positive subjective experience (e.g., well-being and satisfaction, flow, pleasure, happiness, optimism, hope, strengths of character, personal abilities, among many others) (Seligman, 2002; Seligman & Csikszentmihalyi, 2000).

This scientific movement became interdisciplinary and grew immensely and at a fast pace (Marujo & Neto, 2016). Since its inception, positive psychology has been influencing the various areas of psychological intervention, including the clinical arena, conveying the message that psychology is not—and should not be—solely the study of illness, weakness, or decline, being also the study of strengths and virtues (Seligman, 2002).

Positive psychology is therefore a domain that studies mental health and well-being, aiming to investigate how, why, and under what conditions it is possible to encourage the development of positive emotions, positive character, purpose in life, harmonious relationships, and institutions that can assist in this development (Seligman, Steen, Park, & Peterson, 2005). It therefore validated interventions to boost the strengths and virtues that assist individuals to thrive psychologically, socially, and physically in daily life. The exploration of positive mental health, as opposed to the mere absence of mental illness, unlocks salutogenic physiological and psychological shifts and has proved fruitful (Seligman, 2008). Bringing about emotional well-being, in its diverse operationalizations, may be one of our best weapons against mental disorder, which, therefore, may be a relevant buffer against physical disease (e.g., Huffman et al., 2019).

4 Intertwining physical and psychological well-being: A model for positive health

Evidence-based trend analysis from positive psychology data has clearly helped shine light on projected trials and prospects of health in all its scopes. Apart from the uplifting effect on the relevance and consequences of individual and collective psychological flourishing, the relationship between positive mental health and physical health was one of the various axes of research in positive psychology. Martin Seligman called it "positive health," and together with a group of researchers, he steered studies to help classify which particular health resources lead to reduced disease risk and to a lengthier, flourishing life (Seligman, 2008). Research has shown that the scope of these resources might range from biological factors—such as heart rate variability and vagus nerve response—to subjective or functional features, such as purpose in life, optimism, hope, gratitude, a secure and appreciative marital relationship, or a positive leader at work (e.g., Fredrickson, 2013; Kim, Sun, Park, Kubzansky, & Peterson, 2013; Kok & Fredrickson, 2010; Vaillant, 2002). The most important theme that runs through the exciting positive physical health aftermaths is a link between positive psychology and positive health, a truly cross-disciplinary work.

The physiologic progresses associated with positive psychology-based habits are extensive (Lianov et al., 2020), leading to positive global health. Positive global health, as opposed to mere absence of physical illness, has long been ignored scientifically. The discipline of positive health dissolves the dichotomous thinking of mental versus physical salutogenesis. The concept challenges the illness ideology, an ideology that has been socially constructed rather than scientifically raised. This belief consists of not only a set of conventions about the nature of psychological and physical adjustment and the terrain of the health sciences but also a semantic that impacts the way health practitioners and others contemplate the topic. It then proposes a proclamation of a new mission for, and vision for, health sciences and health education, grounded on the values of positive psychology.

According to Martin Seligman, positive health comprehends the consideration that "people desire well-being in its own right and they desire it above and beyond the relief of their suffering. It pronounces a state beyond the mere nonexistence of disease and is definable and quantifiable. Positive health can be operationalized by a blend of outstanding status on biological, subjective, and functional measures" (Seligman, 2008, p. 3). Hence, positive health is not only proper in its own right, but it is also a likely buffer against physical and mental illness, since specific variables that mediate life span, mental health, lower health costs, and better prognosis in specific disorders have already been confirmed by science (Parkes et al., 2020; Seligman, 2008). It forecasts improved longevity (adjusting for quality of life), reduced health expenses, improved mental health in aged people, and superior scenarios when illness appears (e.g., Cohen & Pressman, 2006; Maruta, Colligan, Malinchoc, & Offord, 2000). Increasing evidence has also established that psychological well-being is prospectively linked with superior physical health, even after accounting for baseline health status (Kubzansky et al., 2018; Moskowitz, Addington, & Cheung, 2019; Moskowitz, Epel, & Acree, 2008; VanderWeele et al., 2020).

To exemplify, below there is a list of selected outcomes, among many, from studies in the area of positive health, particularly referenced in the Positive Health Project, supported by the Robert Wood Johnson Foundation, as well as cited in papers by Lianov et al. (2020), Seligman (2008), Kubzansky, Koenen, Spiro, Vokonas, and Sparrow (2007), Kubzansky, Winning, and Kawachi (2014), Kubzansky et al. (2018), and Zammuto, Ottaviani, Laghi, and Lonigro (2021). They all show a clear and positive relationship, on different grounds, between mental and physical health:

- Whether healthy or ill, people with higher life satisfaction (sometimes called happiness) visit the doctor less than those that are unhappy, not running away from the need to visit the doctor when they arise, but distinctly taking better care of themselves that prevents too many doctor appointments;
- People who have positive psychological well-being may be more likely to exercise, eat a healthy diet, and avoid smoking, all behaviors that also reduce their risk of cardiovascular disease;
- There are proven links between positive psychological well-being and greater physical activity in healthy individuals, medically ill persons, and post-ACS patients;

- People with higher life satisfaction are likely to be more optimistic, socially engaged, and supported—and manage health problems better;
- Protecting and enhancing early life psychosocial assets lay the foundation for adult cardiovascular health;
- Type A personalities and people who are chronically angry, anxious, or depressed have a higher risk of heart attacks;
- Men and women with high levels of negative emotion were found to be more likely to die prematurely than those with lower levels of negative emotion;
- While low levels of negative emotion or high levels of cardiorespiratory fitness are predictors of long-term survival in men and women, both being fit and being not unhappy provide a strong combined effect;
- A link between positive affect and reduced mortality was found in chronic illnesses such as HIV and diabetes;
- There is a relation between positive affect and lower risk of first acute coronary syndrome or overall mortality in healthy persons;
- A review of more than 200 studies found a connection between positive psychological attributes, such as happiness, optimism and life satisfaction, and a lowered risk of cardiovascular disease;
- Anger, depression, anxiety, and other negative emotional states can damage heart health;
- Laughter may well be the best medicine for high blood pressure, with a study showing happiness is linked to a 10% reduced risk of hypertension;
- Middle-aged adults who are more optimistic about their future tend to have higher serum antioxidant levels than their less optimistic peers;
- Higher levels of optimism are related to lower risk of having a stroke;
- Optimism and positive affect are independently associated with superior cardiac outcomes and greater physical activity;
- Better levels of "good cholesterol" and other markers of heart health were identified in the blood of middle-aged study subjects with a sunny outlook on life/optimism;
- Poor social support has been linked to depression and loneliness and has been shown to alter brain function and increase the risk of alcohol use, cardiovascular disease, depression, and suicide;
- Those with strong social and emotional support are less likely to die than those who lack such relationship;
- Perceived neighborhood social cohesion and physical health may play an important role in protecting against stroke;
- When it comes to preventing a stroke, living in a neighborhood where you have strong social ties and feel comfortable may be as important as not smoking;
- A higher rating of purpose in life was associated with a reduced likelihood of stroke during a 4-year follow-up.

We can conclude, without any doubt, and referring to mounting evidence, that emotional well-being, as measured by positive emotions, meaning, and other subjective features, protects one from physical illness and impacts longevity and that a state of positive health will increase longevity and improve prognosis (Lianov et al., 2020; Seligman, 2008; Sin, Moskowitz, & Whooley, 2015).

Additionally, there is also interesting data in terms of the accuracy of diagnosis associated with positive emotions in physicians. For instance, brief activities that raise positive moods, like giving chocolates to a physician, enhances creative thinking and makes those physicians more accurate and faster to come up with the proper liver diagnosis (Fredrickson, 2001; Isen, 2005).

These results are strong enough to be relevant contents for health students training—simultaneously for their own sake and personal health contribution and as a way for them to contribute to significantly better holistic health of the ones they serve as practitioners.

Therefore, how can we endorse positive health, enhance well-being, and create conditions for human flourishing? The features of positive health, which specifically foresee these effects, ended up objectives for interventions.

5 Positive psychology interventions for positive health education

As briefly presented, research has acknowledged that healthy habits endorse psychological and physiological health and that emotional well-being is critical to conquering overall well-being (Lianov et al., 2020). Habits based in positive psychology offer the potential to improve Lifestyle-Related Diseases (LRDs) and positive health, in general, by promoting emotional well-being, facilitating healthy lifestyles, and leading to physiologic and behavioral modifications that can tackle the key lifestyle-related public health calamities (Lianov et al., 2020; VanderWeele et al., 2020).

The use of positive psychological interventions may be considered a complementary strategy in mental and physical health promotion and treatment (Bolier et al., 2013). Complementary strategies within and outside the health system that promote healthy lifestyles with positive psychology interventions have shown that they can effectively prevent, treat, and even inverse the bulk of chronic disease cargo (Seligman, 2008). Also, recent research has shown that positive psychology interventions can reduce physician burnout and improve well-being (Bazargan-Hejazi, Shirazi, Wang, et al., 2021).

Therefore, educating health specialists to address issues through the lens of positive psychology can be a valuable complementary educational approach to the usual training. Health professionals can find the techniques that emerged from the more than 20 years of research in positive psychology a useful addition to their skills to help clients and, alongside, to support their own well-being. Positive psychology interventions are rigorously researched exercises, which are inexpensive, simple, and well accepted, can be delivered remotely, and require little provider training compared with more intensive existing programs (Huffman et al., 2019). These techniques, such as savoring, optimistic and gratitude interventions, kindness boosters, strength-building activities, meaning-oriented interventions, and appreciative communication approaches, involve activities and processes widely researched and have overwhelming pieces of evidence that suggest their effectiveness and efficacy (Bolier et al., 2013; Lianov et al., 2020).

Below, there is a list of several of these positive interventions.

(a) *Gratitude letter*

Participants write a letter of gratitude thanking a person for an act of kindness; the letter can be shared (when possible), which intensifies the positive consequences for the persons involved and their relationship.

(b) *Gratitude for positive events*

Participants recall three events that led to satisfaction, happiness, or other positive states, daily for a short period, or every week.

(c) *Capitalizing on positive events*

Participants recall three recent positive events, then share the details with others or otherwise record/celebrate the events.

(d) *Using gratitude in daily life*

Participants focus on implementing gratitude-based interventions and skills into daily life.

(e) *Remembering past successes*

Participants recall a prior event in which they experienced success. They then write about the event, the positive feelings evoked, and their contribution to the success.

(f) *Enjoyable and meaningful activities*

Participants complete activities that bring immediate boosts in mood and others that are more deeply meaningful.

(g) *The "Good Life"*

Participants write about an ideal life over the next year in one or more domains, such as health or relationships.

All these seven interventions are based on the proven relationship between positive affect and health (e.g., Cohen & Pressman, 2006; Fredrickson & Branigan, 2005).

(h) *Assessing personal strengths*

When individuals identify their signature strengths (https://www.viacharacter. org/survey/account/register VIA test) and use their highest strengths in a new way in daily life in general, life satisfaction increases and depression decreases, with 6-month follow-up.

(i) *Using strengths in daily life*

Participants focus on implementing strengths-based interventions and skills into daily life.

The psychological process associated with using strengths seems to be an increase in engagement in activities, and the exercise is self-sustaining. Some authors suggest that this engagement-building intervention will improve prognosis in cardiovascular disease (Seligman, 2008).

(j) *Focusing on life meaning*

Participants focus on implementing enjoyment- and values-based activities and skills into daily life.

A strong sense of purpose is related to a lower risk of all-cause mortality after age 50. Individuals who have a clear goal in life—a reason to get up in the morning, something that makes a difference, from their point of view—tend to live longer and be sharper than those who do not (e.g., Cohen, Bavishi, & Rozanski, 2016).

(k) *Performing acts of kindness*

Participants complete three acts of kindness in 1 day, then write about how the acts made them feel.

Kindness has been shown to benefit our emotional well-being and improve overall happiness. Research has found that performing random acts of kindness can also increase our longevity and boost heart health. Performing acts of kindness has measurable impacts on our mental health by increasing the neurotransmitters in the brain that make us feel satisfied and overall good as serotonin and dopamine. Random acts of kindness toward others can also increase oxytocin, which is a hormone that makes us feel connected to each other and that we can trust each other. These three chemicals can have a profound impact on our mood and overall happiness. Being kind can help reduce levels of the stress hormone cortisol as well (e.g., Brown, Consedine, & Magai, 2005).

(l) Participants apply active-constructive responding (ACR) to positive events reported by a partner, a friend, or a coworker. ACR is a positive and enthusiastic way of responding when someone shares good experiences or information.

If the receiver of the good news actively and constructively responds, capitalizing on the experience, it can often provide a boost in well-being to both people involved in the conversation (Lambert et al., 2013). ACR intensifies marital, friendship, and work happiness (Gable & Reis, 2010; Gable, Reis, Impett, & Asher, 2004).

Its relationship with early stages of lung cancer has been proposed (Seligman, 2008).

(m) *Optimistic thinking/explanatory style*

Participants apply explanatory interpretations to good and bad life events so that optimism is boosted.

The relationship between optimistic explanatory style and cardiovascular disease has been confirmed (Buchanan, 1995).

6 In conclusion: A foresight for health education

A new paradigm in the training of health students can help communities and individuals not just to mitigate damage and heal but also to thrive.

Considering the current social, economic, and, particularly, public health context, it is now imperative that we leverage positive health and deepen the relationship between mental and physical health. Consequently, the aim of the future of health education is to enable access to more complex, interdisciplinary, and holistic training, related to lifestyle diseases and global health, that brings forward well-being as a drive and supports itself on the vast data that links mental and physical health.

Education came to be understood as an asset in national and international development, an angle that endorses studies on how education delivers health profits to both the individual and the society (Marujo, 2021; Marujo & Casais, 2021). New teaching contents, inclusive interdisciplinary learning methods, and innovative pedagogies are relevant mechanisms to prepare the current healthcare student for the new challenges being faced, either in terms of what concerns his/her personal mental and physical health, resilience capacity, and of those that he/she will serve while working a practitioner.

As presented, complementary strategies within and outside the health system that promote healthy lifestyles based in positive psychology interventions can effectively prevent, treat, and even reverse the majority of chronic disease burden globally.

Stakeholders who aim to improve health outcomes and reduce costs should consider ways to facilitate integrating positive psychology interventions into health care and healthcare education. Unfortunately, positive psychology research is often siloed in behavioral health backgrounds, which implies that, despite all the data showing the benefits of positive psychology activities in averting and healing psychological and physical conditions, such interventions are not customarily incorporated in the present-day healthcare ecosystem outside of mental health research and clinical settings. However, these interventions are appropriate and significant for all patients (Lianov et al., 2020). Democratizing these opportunities is a major issue for education systems and societies all over the world.

Current evidence on social relations and health, specifically social integration and social support, shows a relationship to health behaviors and health outcomes (Cohen & Janicki-Deverts, 2009). The evidence is also dense regarding the relationship between physical and mental health (Seligman, 2008). Therefore, pursuing the link between positive psychological factors and healthcare use may reveal innovative ways to hold healthcare costs. Positive psychological well-being, desirable in itself, may have far-reaching consequences for physical health (Fredrickson & Branigan, 2005).

Collaborative translational research should adapt the positive psychology and behavioral medicine evidence base into methodologies that address emotional well-being in nonmental health care settings.

Future research should continue to test the scalability of effective positive psychology interventions, tools, and practices in everyday health care and health education. Nevertheless, what we already know should unequivocally promote health system changes in order to integrate evidence-based positive-psychology interventions into health maintenance and treatment plans. Additionally, it should ameliorate health provider well-being through the application of positive psychology methods for providers' personal health.

Investing in research, policy, and practice innovations to promote total well-being via lifestyle changes considers that health sciences education must respond to the question, "What do we want to become?" To think in terms of "becoming" is to invoke a line of thought that emphasizes potentials, rejects determinism, and expresses a flexible openness to innovation (UNESCO, 2021).

The new generations being prepared—notwithstanding any crises—are empowered with the promise of a forthcoming bursting with hope and possibility that embraces a fluid, iterative, rigorous, and collective approach to futures-making.

Positive psychology data and interventions might serve as a sustainable paradigm for organizing public health education and health readiness, response, and recovery. By bolstering health systems, meeting the needs of exposed populations, and upholding individual and organizational capability, social connectedness, and psychological health, positive health supports actions that build awareness, promote strong everyday systems, and tackle the underlying psychological, economic, social, and cultural determinants of health. Thus, positive health resonates with a wide array of stakeholders, particularly those whose work routinely addresses health, wellness, or societal well-being education.

7 Five active learning sections

1. **Suggested teaching assignments:**
   - Write a report on the relationship between mental and physical health
   - Compile data on the new tendencies in health education regarding the inclusion of well-being as a subject
   - Write a draft on the definition and relevance of positive psychology
   - Write a gratitude letter to a colleague, a teacher, or a family member; read it to the person; and look for data regarding its physiological and relational effects
2. **Recommended readings:**
   Articles:
   - Good genes are nice, but joy is better, Harvard Gazette:
     https://news.harvard.edu/gazette/story/2017/04/over-nearly-80-years-harvard-study-has-been-showing-how-to-live-a-healthy-and-happy-life/

- Harvard Study of Adult Development:
  https://www.adultdevelopmentstudy.org/
  Videos:
- Fredrickson, B. How positive emotions work and why (2019).
  https://www.youtube.com/watch?v=nD_SbilNMo4
- Seligman, M. Positive Psychology in a Pandemic. (2021).
  https://www.youtube.com/watch?v=L1hauE_OKP8
- 19 Best Positive Psychology Interventions and why to apply them:
  https://positivepsychology.com/positive-psychology-interventions/

3. Case study

When we work as health practitioners, there is a lot of potential for things to go wrong: the procedures are quite complex, the clients are almost always in vulnerable positions, and the practitioner has a huge responsibility and very high ethics standards and is frequently emotionally exposed. While the work context for these healthcare workers is already and usually convoluted, recently, the COVID-19 pandemic has had a particular impact on health professionals' psychological health. Although the well-being and emotional resilience of healthcare professionals are key components of continuing healthcare services, these professionals have been observed in this period to experience serious psychological problems and to be at risk in terms of mental health. Healthcare workers constitute the most affected group of people in the fight against the COVID-19 virus. Among the common mental effects of the pandemic are anxiety, panic, depression, anger, confusion, ambivalence, and financial stress. Studies have revealed that healthcare workers are exposed to work overload, isolation, and discrimination, and therefore they experience exhaustion, fear, affective disorders, and sleep problems. Healthcare workers were observed to experience similar problems during previous pandemics, namely during the 2003 SARS and 2014 Ebola virus pandemics. Studies have also shown that healthcare professionals are considerably more worried about catching the infection during a pandemic and they might become traumatized; particularly, those working in public health, primary care, emergency service, and intensive care are at risk of developing psychological symptoms. Hence, mood management is required to avoid exacerbation of psychological symptoms and burnout, and being resilient is therefore vitally important for the healthcare practitioner, in order to raise psychological resilience, positive emotions, life satisfaction, and quality of sleep.

Taking this into consideration and what the positive psychology and positive health research has shown so far, please answer the following questions:

Q1: What is psychological resilience?

Q2: Why is it particularly important for the holistic health of healthcare professionals?

Q3: Will low levels of psychological resilience, positive emotions, and well-being potentially interfere with the quality of professional practices? If yes, what are the risks?

Q4: Can meaning and purpose, when attached to the work they provide, as well as frequent positive emotions, be psychological buffers for physical and emotional health perils?

Q5: If you had to plan training on psychological resilience for healthcare practitioners, what contents would you include?

4. **Titles for research essays**
   - Positive psychological well-being and cardiovascular disease
   - Psychological Health, Well-Being, and the Mind-Body Connection
   - Longevity and optimism
   - How positive emotions build physical health
   - Negative affect and physical symptoms
   - Gratitude interventions: effects on physical health

5. **Recommended Projects URL (indicative URLs from current initiatives, professional websites, or project URLs related to your discussion)**
   - Exploring the Concept of Positive Health

      Building on advances in Positive Psychology, the emerging field of Positive Health is examining whether factors such as optimism and happiness may lead to better health and well-being. Publisher: Robert Wood Johnson Foundation.

      URL: https://www.rwjf.org/en/library/research/2017/08/exploring-the-concept-of-positive-health.html
   - Positive Health: Researching strengths that contribute to good health and protect against illness

      URL: https://positivehealthresearch.org/

References

Alsoufi, A., Alsuyihili, A., Msherghi, A., Elhadi, A., Atiyah, H., Ashini, A., et al. (2020). Impact of the COVID-19 pandemic on medical education: Medical Student's knowledge, attitudes, and practices regarding electronic learning. *PLoS ONE*, *15*(11), e0242905. https://doi.org/10.1371/journal.pone.0242905. 33237962. eCollection 2020.

Bazargan-Hejazi, S., Shirazi, A., Wang, A., et al. (2021). Contribution of a positive psychology-based conceptual framework in reducing physician burnout and improving well-being: A systematic review. *BMC Medical Education*, *21*, 593. https://doi.org/10.1186/s12909-021-03021-y.

Bolier, L., Haverman, M., Westerhof, G. J., Riper, H., Smit, F., & Bohlmeijer, E. (2013). Positive psychology interventions: A meta-analysis of randomized controlled studies. *BMC Public Health*, *13*, 119. https://doi.org/10.1186/1471-2458-13-119.

Brown, W. M., Consedine, N. S., & Magai, C. (2005). Altruism relates to health in an ethnically diverse sample of older adults. *The Journals of Gerontology: Series B: Psychological Sciences and Social Sciences*, *60*(3), 143–152.

Buchanan, G. M. (1995). Explanatory style and coronary heart disease. In G. M. Buchanan, & M. E. P. Seligman (Eds.), *Explanatory style* (pp. 225–232). Hillsdale, NJ: Erlbaum.

Carvalho, Y. M., & Ceccim, R. B. (2012). Formação e educação em saúde: Aprendizagens com a saúde coletiva. G. W. S. Campos, M. C. S. Minayo, M. Akerman, M. Drumond Júnior, & Y. M. Carvalho (Org.) In *Tratado de Saúde Coletiva* (2nd ed.). São Paulo: Hucitec.

Cohen, R., Bavishi, C., & Rozanski, A. (2016). Purpose in life and its relationship to all-cause mortality and cardiovascular events: A meta-analysis. *Psychosomatic Medicine, 78*(2), 122–133.

Cohen, S., & Janicki-Deverts, D. (2009). Can we improve our physical health by altering our social networks? *Perspectives on Psychological Science, 4*(4), 375–378.

Cohen, S., & Pressman, S. D. (2006). Positive affect and health. *Current Directions in Psychological Science, 15*, 122–125.

Costa, M., & López, E. (2008). La perspective de la potenciación en la intervención psicológica. In Vásquez, & G. Hervás (Eds.), *Psicología positiva aplicada*. Espanha: Descleé de Brouwer.

Faerber, A., Andrews, A., Lobb, A., Wadsworth, E., Milligan, K., Shumsky, R., et al. (2019). A new model of online health care delivery science education for mid-career health care professionals. *Healthcare, 7*(4). https://doi.org/10.1016/j.hjdsi.2018.12.002.

Fredrickson, B. L. (2001). The role of positive emotions in positive psychology: The broaden-and-build theory of positive emotions. *American Psychologist, 56*(3), 218–226. https://doi.org/10.1037/0003-066X.56.3.218.

Fredrickson, B. L. (2013). Positive emotions broaden and build. *Advances in Experimental Social Psychology, 47*, 1–53.

Fredrickson, B. L., & Branigan, C. A. (2005). Positive emotions broaden the scope of attention and thought–action repertoires. *Cognition and Emotion, 19*, 313–332.

Freudenberg, N. (1978). Shaping the future of health education: From behavior change to social change. *Health Education Monographs, 6*(4), 372–377.

Gable, S. L., & Reis, H. T. (2010). Good news! Capitalizing on positive events in an interpersonal context. In *Vol. 42. Advances in experimental social psychology* (pp. 195–257).

Gable, S. L., Reis, H. T., Impett, E. A., & Asher, E. R. (2004). What do you do when things go right? The intrapersonal and interpersonal benefits of sharing positive events. *Journal of Personality and Social Psychology, 87*, 228–245.

Gebbie, K., Rosenstock, L., & Hernandez, L. M. (Eds.). (2003). *Who will keep the public healthy? Educating public health professionals for the 21st century*. Washington, DC: National Academies Press (US). The Future of Public Health Education. Institute of Medicine (US) Committee on Educating Public Health Professionals for the 21st Century. Available on 15 November 2021 from: https://www.ncbi.nlm.nih.gov/books/NBK221190/.

Huffman, J. C., Feig, E. H., Millstein, R. A., Freedman, M., Healy, B. C., Chung, W., et al. (2019). Usefulness of a positive psychology-motivational interviewing intervention to promote positive affect and physical activity after an acute coronary syndrome. *The American Journal of Cardiology, 123*(12), 1906–1914.

IOM (Institute of Medicine). (2011). *The future of nursing: Leading change, advancing health*. Washington, DC: The National Academies Press.

Isen, A. M. (2005). A role for neuropsychology in understanding the facilitating influence of positive affect on social behavior and cognitive processes. In C. R. Snyder, & S. J. Lopez (Eds.), *Handbook of positive psychology* (pp. 528–540). Oxford: Oxford University Press.

Jackson, S. (2000). Joy, fun, and flow state in sport. In Y. Hanin (Ed.), *Emotions in sport (cap. 6)*. Champaign: Human Kinetics.

Jourdan, D., Samdal, O., Diagne, F., & Carvalho, G. S. (2008). The future of health promotion in schools goes through the strengthening of teacher training at a global level. *Promotion & Education, 15*(3), 36–38.

Kim, E. S., Sun, J. K., Park, N., Kubzansky, L. D., & Peterson, C. (2013). Purpose in life and reduced risk of myocardial infarction among older U.S. adults with coronary heart disease: A two-year follow-up. *Journal of Behavioral Medicine, 36*(2), 124–133.

Kok, B. E., & Fredrickson, B. L. (2010). Upward spirals of the heart: Autonomic flexibility, as indexed by vagal tone, reciprocally and prospectively predicts positive emotions and social connectedness. *Biological Psychology, 85*(3), 432–436.

Kubzansky, L. D., Huffman, J. C., Boehm, J. K., Hernandez, R., Kim, E. S., Koga, H. K., et al. (2018). Positive psychological well-being and cardiovascular disease. *Journal of the American College of Cardiology, 72*, 1382–1396.

Kubzansky, L. D., Koenen, K. C., Spiro, A., Vokonas, P. S., & Sparrow, D. (2007). A distressed heart: A prospective study of posttraumatic stress disorder symptoms and coronary heart disease. *Archives of General Psychiatry*, *64*, 109–116.

Kubzansky, L. D., Winning, A., & Kawachi, I. (2014). Affective states and health. In L. F. Berkman, M. M. Glymour, & I. Kawachi (Eds.), *Social epidemiology: New perspectives on social determinants of global population health* (2nd ed.). New York: Oxford University Press.

Lambert, N. M., Gwinn, A. M., Baumeister, R. F., Strachman, A., Washburn, I. J., Gable, S. L., et al. (2013). A boost of positive affect: The perks of sharing positive experiences. *Journal of Social and Personal Relationships*, *30*(1), 24–43.

Lianov, L. S., Barron, G. C., Fredrickson, B. L., Hashmi, S., Klemes, A., Krishnaswami, J., et al. (2020). Positive psychology in health care: Defining key stakeholders and their roles. *Translational Behavioral Medicine*, *10*(3), 637–647.

Liu, Q., et al. (2016). The effectiveness of blended learning in health professions: Systematic review and meta-analysis. *Journal of Medical Internet Research*, *18*(1). https://doi.org/10.2196/jmir.4807.

Lucey, C. R., Thibault, G. E., & Ten Cate, O. (2018). Competency-based, time-variable education in the health professions: Crossroads. *Academic Medicine*, *93*(3), S1–S5. https://doi.org/10.1097/ACM.0000000000002080.

Marujo, H.Á. (2021). O perigo das ameaças à paz: Uma ode à justiça global. Fauston Negreiros & Jorge Rio Cardoso (Coord.) In *Reflexões sobre Psicologia e Educação no contexto da pandemia no Brasil e em Portugal*. Teresina: EDUFPI, Instituto Benjamin Franklin.

Marujo, H.Á., & Casais, M. (2021). Educating for public happiness and global peace: Contributions from a Portuguese UNESCO chair towards the sustainable development goals. *Sustainability*, *13*(16), 9418. https://doi.org/10.3390/su13169418.

Marujo, H.Á., & Neto, L. M. (2016). Quality of life studies and positive psychology. In L. Bruni, & P. L. Porta (Eds.), *Handbook of research methods and applications in happiness and quality of life* (pp. 279–305). Cheltenham, UK: Edward Elgar Publishing.

Maruta, T., Colligan, R. C., Malinchoc, M., & Offord, K. P. (2000). Optimists vs. pessimists: Survival rate among medical patients over a 30-year period. *Mayo Clinic Proceedings*, *75*, 140–143.

Moskowitz, J. T., Addington, E. L., & Cheung, E. O. (2019). Positive psychology and health: Well-being interventions in the context of illness. *General Hospital Psychiatry*, *61*, 136–138. https://doi.org/10.1016/j.genhosppsych.2019.11.001.

Moskowitz, J. T., Epel, E. S., & Acree, M. (2008). Positive affect uniquely predicts lower risk of mortality in people with diabetes. *Health Psychology*, *27*, S73–S82.

Neves, A. C. (2021). An amplified concept of health. *Academia Letters*, 3872. https://doi.org/10.20935/AL387.

OECD. (2018). *The future of education and skills. Education 2030: The future we want*. Retrieved from https://www.oecd.org/education/2030/E2030%20Position%20Paper%20. (Accessed 3 October 2021).

Parkes, M. W., Poland, B., Allison, S., et al. (2020). Preparing for the future of public health: Ecological determinants of health and the call for an eco-social approach to public health education. *Canadian Journal of Public Health*, *111*, 60–64. https://doi.org/10.17269/s41997-019-00263-8.

Saliceti, F. (2015). Educate for creativity: New educational strategies. In *7th world conference on educational sciences*. Athens, Greece: Science Direct.

Seligman, M. (2002). Positive psychology, positive prevention, and positive therapy. In C. Snyder, & S. Lopez (Eds.), *Handbook of positive psychology (cap.1)*. New York: Oxford University.

Seligman, M. E. P. (2008). Positive health. *Applied Psychology. An International Review*, *57*(Suppl 1), 3–18.

Seligman, M., & Csikszentmihalyi, M. (2000). Positive psychology: An introduction. In M. Seligman, & M. Csikszentmihalyi (Eds.), *American Psychologist: Vol. 55(1). Special issues on happiness, excellence, and optimal human functioning* (pp. 5–14).

Seligman, M. P. E., Ernst, R. M., Gillham, J., Reivich, K., & Linkins, M. (2009). Positive education: Positive psychology and classroom interventions. *Oxford Review of Education*, *35*(3), 293–311. https://doi.org/10.1080/03054980902934563.

Seligman, M., Steen, A., Park, N., & Peterson, N. (2005). Positive psychology progress—Empirical validation of interventions. *American Psychologist*, *60*(5), 410–421.

Sin, N. L., Moskowitz, J. T., & Whooley, M. A. (2015). Positive affect and health behaviors across 5 years in patients with coronary heart disease: The Heart and Soul Study. *Psychosomatic Medicine, 77*, 1058–1066.

UN General Assembly. (2015). *Transforming our world: The 2030 agenda for sustainable development.* A/RES/70/1, Retrieved from https://www.refworld.org/docid/57b6e3e44.html. (Accessed 1 October 2021).

UNESCO. (2021). *Futures of Education Initiative.* Retrieved from https://en.unesco.org/futuresofeducation/initiative. (Accessed 27 November 2021).

United Nations Children's Fund (UNICEF). (2021). *Ensuring equal access to education in future crises: Findings of the new remote learning readiness index.* New York: UNICEF.

Vaillant, G. E. (2002). *Aging well: Surprising guideposts to a happier life from the landmark Harvard Study of Adult Development.* New York: Little, Brown.

VanderWeele, T. J., Trudel-Fitzgerald, C., Allin, P., Farrelly, C., Fletcher, G., Frederick, D. E., et al. (2020). Current recommendations on the selection of measures for well-being. *Preventive Medicine, 133*, 106004.

World Health Organization. (1946). Preamble to the Constitution of the World Health Organization. *Official Records of the World Health Organization, 2*, 100.

Zammuto, M., Ottaviani, C., Laghi, F., & Lonigro, A. (2021). The heart in the mind: A systematic review and meta-analysis of the association between theory of mind and cardiac vagal tone. *Frontiers in Physiology, 12*, 611609. https://doi.org/10.3389/fphys.2021.611609. 34305625. PMCID: PMC8299530.

Further reading

Cohen, G. D., Perlstein, S., Chapline, J., Kelly, J., Firth, K. M., & Simmens, S. (2006). The impact of professionally conducted cultural programs on the physical health, mental health, and social functioning of older adults. *Gerontologist, 46*(6), 726–734.

Huffman, J. C., DuBois, C. M., Mastromauro, C. A., Moore, S. V., Suarez, L., & Park, E. R. (2016). Positive psychological states and health behaviors in acute coronary syndrome patients: A qualitative study. *Journal of Health Psychology, 21*, 1026–1036.

Kubzhansky, L. D., & Thurston, R. (2007). Emotional vitality and incident coronary heart disease. *Archives of General Psychiatry, 64*, 1393–1401.

Positive Heath Project. Retrieved on 11th of October 20210 from https://positivehealthresearch.org/.

VanderWeele, T. J. (2020). Challenges estimating total lives lost in COVID-19 decisions: Consideration of mortality related to unemployment, social isolation, and depression. *JAMA.* https://doi.org/10.1001/jama.2020.12187.

CHAPTER 11

Public health communication for seniors: *E-health*, enabling and empowerment

Rita Espanha[a] and Francisco Garcia[b]
[a]Communication, Culture and Information Technologies, Iscte-University Institute of Lisbon, Lisbon, Portugal
[b]Communication Sciences, Iscte-University Institute of Lisbon, Lisbon, Portugal

1 Present-day communication

The intense diffusion and circulation of information is one of the key features that characterize modern Western societies. The increased distribution and access to information and the themes and forms in which the democratization of knowledge and information takes place are multiple, diverse, and broad.

At the turn of the millennium, Giddens (2000) affirmed that in current societies, the relationships and interrelationships between individuals and institutions as well as systems and organizations would diversify, intensify, and become more complex, as access to information and communication arises from the evolution of information and communication technologies (ICT) and their growing presence in people's lives.

But can it be said that the media currently reinforces the ontological security (Giddens, 2000) provided to us by the content and its format (Silverstone, 1999)? Are the media central in the social construction of reality, as described by Berger and Luckmann (2004)? Or should media discourses be considered active participants in the construction of reality, thus creating social representations, by linking current to previous phenomena, revealing what was only imagined, and making concrete what was theoretical (Moscovici, 2005)?

Espanha (2020, p. 327) emphasizes that "the assumption is that social representations shape what is provided to us from outside, from the relationship of individuals and groups with objects, acts and situations that are created in countless social interactions; based on this, the reproduction made by this representation implies changes in structures and elements, that is, a reconstruction of what is given in the context of values, rules and concepts".

She further warns that the understanding that there is no rupture between the concepts of "outer universe" and individual "collective universe" is pivotal because, based on the ideas of Moscovici (1978): "The object is inserted in a dynamic context, partially

Active Learning for Digital Transformation in Healthcare Education, Training and Research
https://doi.org/10.1016/B978-0-443-15248-1.00017-5

conceived by the collective or by the individual as an extension of their behaviour" (pp. 327–328).

In this context, communication manifests as a transaction or transmission of information. This in turn implies the existence of a dynamic of sharing between the actors involved in the process, "being able to make needs known, exchange information, ideas, attitudes and beliefs, create understandings, establish and maintain relationships" (p. 328).

2 Health communication and challenges in aging societies

Understanding the theoretical and practical components of communication in health requires us to reflect on the literal meaning of the word communication: "a process by which information is exchanged between individuals through a common system of symbols, signs, or behavior; personal rapport; information communicated: information transmitted or conveyed; a verbal or written message; a system (as of telephones or computers) for transmitting or exchanging information ..." (Merriam-Webster Dictionary, Oct., 2021).[a]

Schiavo (2014) argues that these definitions are essential in the design of models for health communication interventions. She emphasizes that, as in other forms of communication, health communication must be based on a logic of bidirectional information exchange with a common system of behavior and signals between the parties. He also states that it must be accessible and capable of creating mutual feelings of understanding and empathy between those who communicate, audiences, and key groups, that is, "all audiences the health communication program is seeking to influence and engage in the communication process" (p. 4).

In the view of the abovementioned author, the term "key group" occupies a special place in health communication, as it helps to define the participatory nature of an intervention of this type more precisely, given that communities and other key groups assume the role of agents of social change (Schiavo, 2014). Indeed, considering that the concept of audience can have a more passive connotation, coining the term "key group" indicates the importance of creating these groups, in that they help to better define the priorities, needs, and preferences to which a health communication plan should attend.

Along the same line, Espanha (2019) defines health communication as "the study of communication strategies to inform and influence individual and collective decisions involving health issues and the promotion of autonomy, necessarily linking the domains of health and communication" (p. 15).

The author explains that this branch of communications can be defined as a field of studies which has different scientific influences from the social and health sciences. "It is a field that, while it is more developed in countries like the USA or Brazil, is still under

---

[a] https://www.merriam-webster.com/dictionary/communication.

construction; it is of a multidimensional nature and is where individuals and institutions develop strategies and weave alliances, antagonisms and negotiations" (Espanha, 2019, p. 15). This concept implies bringing to the forefront the existence of competing discourses, which are constituted by and constitute relations of knowledge and power—a dynamic that, in its turn, encompasses the different spectrums of communication, health, and the relationships between them (Araújo & Cardoso, 2007; Espanha, 2019).

Health communication has thus gradually cemented its place in various (practical and theoretical) aspects of public health, namely in health promotion, environmental health, health policies, and global health (Abroms & Maibach, 2008; Bernhardt, 2004; Kreps, 2001; Kreps & Maibach, 2008). From the same perspective, Schiavo (2014) states that communication in health plays an extremely important role as it aims to interact with, empower, and influence individuals and the communities in which they are located.

The author asserts that communication in health constitutes an admirable cause, insofar as it seeks to improve the state of public health through the sharing of information on health and introduces us to the concepts of "vulnerable populations", "underserved populations", and "health equity" (Schiavo, 2014, p. 6).

In this case, "vulnerable population" is defined as the groups of the population that are at greatest risk in terms of showing signs of physical, psychological, and/or social weakness and living in situations in which there is a lack of adequate conditions to achieve stability in life (for example, children, seniors, people with disabilities, migrant populations, or groups affected by stigma and social discrimination). As regards "underserved populations", these are considered to be specific groups (ethnic, social and geographical) in communities that do not have means of access to services and infrastructure or information. Finally, "health equity" is defined as the act of making it possible for all citizens to have equal opportunities for access to health care, being able to overcome diseases and health crises—thus overcoming the barriers of ethnicity, gender, age, economic status, social status, and environment, among other factors.

In short, equity can only be achieved via the creation of a receptive, favorable environment in which information can be properly shared, perceived, absorbed, and discussed by different communities and sectors in order to be inclusive and representative of the circumstances of vulnerable and disadvantaged populations (Schiavo, 2014).

Berry (2007) points out that, particularly in the Western situation, seniors (aged 65 years or over) are one of the most challenging groups for health communication, as their numbers are growing due to the increase in average life expectancy. The challenge for this author lies in the attempt to reconcile meeting the needs and demands of seniors toward the health systems with bearing the weight inherent in this effort—"we know, for example, that people in this age group are more likely to be taking prescribed medication than their younger counterparts" (p. 52).

As they are also at greater risk of developing multiple pathologies, many of which are chronic in nature and require long-term medication, seniors tend to trust and rely heavily on medical professionals. Communication between parties may thus be characterized by certain factors or "typical communication patterns" that result from the patient/caregiver interaction (Berry, 2007, p. 53).

Berry (2007) also emphasizes that the treatment given by doctors to seniors may become condescending, brusque, indifferent, and careless. Attitudes such as speaking loudly and paternalistically are often seen, treating seniors as if they were children and not adults to the full extent of their abilities.

With regard to senior patients, the process of communication is less effective if they are withdrawn and do not, for example, report a lot of information, or if they are not assertive and do not ask health professionals many questions. In certain situations, this type of communication barrier can also arise from physical and/or cognitive factors, which can add to the aforementioned behavioral factors.

A third dynamic that can hinder the communication process occurs in situations where seniors are accompanied to an appointment by their partner, for example, or by one of their relatives. In this circumstance, there is a high degree of probability that health professionals will be tempted to talk directly to the caregiver and not the patient, particularly if the caregiver is younger (Berry, 2007).

However, Marques (2011) warns that discrimination against seniors, or ageism—a term first coined in 1969 by psychologist Robert Butler—may increase as the population grows increasingly older. He tells us that "in general terms, ageism refers to generalized negative attitudes and practices towards individuals based on only one characteristic – their age" (p. 18). Ageism can occur between different age groups, but in relation to older people, it can be identified in the following three aspects, in the author's view (Marques, 2011):

(1) Related to beliefs or stereotypes that exist about seniors, under the assumption that all people over a certain age are part of a homogeneous group, often connoted with certain negative aspects (such as disability or illness).

(2) Ageist attitudes are closely related to prejudice and feelings that we may have toward the senior age group, for example, which are manifested through feelings of disdain. On some occasions, these feelings may materialize in the form of pity, or in paternalistic behavior.

(3) Finally, ageism can also include a behavioral facet, which can be seen in effective acts of discrimination committed against seniors. A possible example lies in situations in which cases of abuse and mistreatment occur.

Marques (2011) states that ageism goes beyond the characteristic of an individualized negative attitude toward seniors and that the concept particularly projects "our deepest cultural values and institutional practices of our society" (p. 19).

Indeed, Marques et al. (2020) emphasize that ageism is one of the greatest threats to active aging, as it is manifested in a wide range of domains ranging from individual to institutional and cultural age components. Rebelo (2016) adds that in a society where digital technologies are increasingly present and widespread, we must understand that "being digitally excluded represents a situation of great inequality and social disadvantage for individuals" (p. 154)—a problem that has special relevance in the Portuguese scenario, due to the proportion of people who still do not have full access to participation in the digital sphere.

The component of political and social participation in the lives of seniors, in addition to family and community contexts, is essential. Despite attempts to promote initiatives aimed at promoting active aging for the exercise of full citizenship, prejudice and ageism continue to be spread in societal discourse.

Rebelo also emphasizes that "these negative discourses, ideas and attitudes about old age may be at the origin of the idea that technology is not for the elderly or that it is too complicated for them" (p. 154) and that it is for these reasons that older people have a lack of interest in and motivation for using digital technologies and often avoid them (Dias, 2012; Lüders & Brandtzæg, 2014; Lugano & Peltonen, 2012; Morris, Goodman, & Branding, 2007; Selwyn, 2012).

3 *E-health* + literacy = more able and included seniors?

The term E-health has no consensual definition within the academic community. Eysenbach (2001) alone identified 51 definitions for "*E*-health" based on the scientific literature, establishing a link between the concepts of health and technologies.

E-health can therefore be seen as an area that has been consolidated by the joining of forces between medical informatics, public health, and the business context. It operates in the field of health and information services and is communicated using the Internet and other information and communication technologies (ICT) (Espanha, 2019; Eysenbach, 2001; Schiavo, 2014).

Espanha (2019) explains that E-health aims to establish a new form of relationship between citizens and health professionals, based on sharing information and decisions— "However, there is a long, urgent path to take in this direction, because for this to become reality, it is important to make relevant scientific information available, by reliable, quality and safe electronic means, as well as the personal health records of citizens" (p. 44).

From this perspective, E-health is defined by the health information and communication networks, which are online, aimed at the general public and health professionals, and constructed in this context. Also included is the provision of all kinds of information services, construction of platforms, availability of content, and digital registration of patients/users, also available online so that they can be consulted and used by all Internet users (directly or indirectly). According to this definition, information networks directed to the health area and online portals for health, promotion, and provision of services and/ or remote care should also be included.

Hou (2012) constructed a model with six items that he says are essential in conceptualizing a more accessible health website for all types of Internet users, particularly those with low levels of literacy, listed as follows: "(1) the presentation of information should be made taking into account the description of health behavior and the benefits of taking action and taking specific action steps; (2) text should be presented in a way that motivates action and that is not limited to describing, including placing the most important information in the forefront: describing the type of behavior to adopt, encouraging users to remain positive and realistic, providing a list of specific actions that can be performed by steps; writing in clear language, and making sure the information is correct through fact checking; (3) a strategy should be used to display the contents on the web page intuitively; (4) content should be organized and navigation simplified; (5) users should be attracted with interactive content, for example, giving them a hyperlink that allows them to print the pages, as many users still prefer to read on paper, especially people with low literacy levels; (6) website content should be frequently reviewed and revised. It is recommended that this ongoing process involve groups of users with low levels of health literacy and groups of experienced moderators. This is done to test the levels of understanding of the exposed content, consider the type of interaction between the different groups and the level of self-efficacy they present, and finally create test documents with clear language.

Considering the most vulnerable populations, particularly seniors, Noblin and Rutherford (2017) conclude that the use of information and communication technologies (ICT) in health care is extremely important and enhances healthy aging. The authors argue that if it can be ensured that senior adults are able to understand the health information provided to them, it may have a direct and positive impact on their quality of life.

Sak et al. (2017) confirm the existence of a positive link between patients being involved in the decision-making process and improved health outcomes. They emphasize that the empowerment and health literacy of older adults influences the willingness of individuals to take a more active role in health-related decisions and that public efforts in health should be made toward developing programs and providing appropriate/adapted information to facilitate this process, especially for individuals with moderate levels of health literacy.

Furthermore, Parker and Ratzan (2019) emphasize that health literacy is fundamental in democratizing health. They add that there are complexities and requirements in aligning health care with the skills of individuals to access, navigate, understand, and use information they need to consult—"Transparency and trustworthiness should characterize not only what is communicated but also how essential information is used and shared" (p. 952).

Regarding digitally including seniors during the pandemic resulting from COVID-19, Afonso, Fernandes, and Magalhães (2020) state that there have been major changes in the paradigm of communication, provision of health services, and active aging.

According to the authors, digital inclusion has proved to be one of the most comprehensive tools for strengthening social inclusion and combating the generational "digital divide" and is a vector of independence for seniors as it increases their capacities/possibilities for social participation.

However, along these lines, the authors also present the case of *Portugal Digital: Apresentação do Plano de Ação para a Transição Digital* (Digital Portugal: Presentation of the Action Plan for the Digital Transition) (Diário da República, 2020), and criticize it heavily for the lack of initiative in the digital inclusion of seniors—"At present, is digital inclusion of the elderly a priority on the agenda in Portugal? (...) the answer leans to the negative, since there is no mention of the elderly population, the urgent need to eliminate the digital divide that separates the young from the elderly or construction of educational models which are adapted to the situation and needs of the elderly" (Afonso et al., 2020, p. 136).

Nevertheless, this document does mention a program aimed at the digital inclusion of 1 million adults, but it does not mention information on the age groups on which the program focuses – let us not forget that the current Portuguese demographic situation points to 2.2 million people aged 65 or older and the trend, according to official data from INE (Instituto Nacional da Estatística—Statistics Portugal),[b] indicates progressive growth, reaching 3 million seniors by 2080.

In the view of the abovementioned authors, senior universities have gradually assumed an important role in promoting the digital inclusion of seniors, with training programs/actions on ICT, often via partnerships and protocols with public and private entities, which create the ideal conditions for this purpose. In fact, during the COVID-19 pandemic, attempts at digital inclusion were bolstered through distance learning models using information and communication technologies—senior universities were thus able to proceed "with their mission to promote of active ageing and encouragement of the social participation of the elderly" (Afonso et al., 2020, p. 137).

With the outlook of a possible intervention with the aim of increasing digital inclusion rates, the concept of "positive technology", coined by Riva et al. (2016), is fundamental as it promotes a scientific approach to the use of technology in achieving improved quality of life, experiences, and well-being of seniors. The concept implies that when using technology, seniors feel supported in their social context and in their interpersonal relationships while not neglecting the element of personal contact that can be more genuine and generate positive affect.

Finally, the authors emphasize that the use and promotion of technology in the gerontological context must be conducted in a rational, ethical, and balanced manner while not neglecting the human component (Riva et al., 2016).

[b] https://www.ine.pt/xportal/xmain?xpid=INE&xpgid=ine_destaques&DESTAQUESdest_boui=406534255&DESTAQUESmodo=2&xlang=pt.

4 Empowerment and enabling of seniors: A path to be followed

In general, in the present day, the question can be posed of the most correct definition for the concept of empowerment. At the turn of the millennium, Rowlands (1995) asserted that a possible definition for the concept could be reached by analyzing the relationship between an individual's ability to interpret the broad concept of "power" and his/her inclusion in the decision-making process—"This puts a strong emphasis on access to political structures and formal decision-making and, in the economic sphere, on access to markets and incomes that enable people to participate in economic decision-making" (p. 102).

Likewise, empowerment can be equally important in decision-making in health matters, as "health literacy has become increasingly central in reflections on health systems and, above all, in the way individuals interact with these systems, in all their current dimensions" (Espanha et al., 2015, p. 74).

Gamliel and Gabay (2014) reinforce the importance of computers and the Internet in this context. They affirm that they represent technological tools and sociocultural platforms for interaction between generations, exchange of knowledge, and empowerment. The results of their study demonstrate that there is a pattern of relationship between empowerment, knowledge exchange, and social interaction (although each generational group analyzed presents different markers).

The research carried out in this area also shows that seniors are willing to get involved in activities that involve the use of computers and can attain significantly positive learning results. This is particularly the case when they participate in short-term intergenerational programs that allow them a proactive role in the process of learning to use computers, which also contributes indirectly to stimulating their self-sufficiency and social involvement.

Sales et al. (2009), in a study carried out with groups of seniors, demonstrate that the peer learning model is beneficial for learning digital skills. They emphasize that digital inclusion can play a fundamental role in strengthening seniors' self-esteem and make them more independent and capable of using technologies. In this context, communication is presented as a key factor for the maintenance and expansion of the seniors' social circle—factors such as this "justify the importance to create interaction alternatives to insert the elderly person in activities such as the use of computers and its communication and information tools, that stimulate, integrate and can expand their life goals, also approaching them to similar technologies, like cell phones and ATM machines, making their utilization easier" (p. 438).

Hill et al. (2015), in turn, claim that senior adults in their study sample recognize the value of technology as a tool for empowerment that can not only facilitate their daily activities but also help maintain social relationships, since it has the power to break down physical and geographic barriers associated with aging.

Although several positive aspects of digitization were registered, some seniors stated that technology can sometimes have the opposite effect and, therefore, be "disempowering". In particular, it was concluded that without an appropriate skill set and without measures to combat the fear associated with technology, the digital divide could widen as more services migrate to the virtual world.

In addition, it is likely that the digital divide may also increase the rate of social isolation and thus reduce access to key services, given the ways in which society and the business world are migrating to an exclusively online state. Consequently, the impact of technology at the level of inclusion (from a micro and macro perspective) should bring about a reflection on the policies which are in place to fight the barriers that separate seniors from technology.

Within this outlook, Wildemeersch and Jütte (2017) highlight the idea that technology is clearly present in our daily practices (for example, commercial activities; banking transactions; entertainment and leisure; communication in general; listening to music; watching television and playing video games), and more precisely, that digitization in work has had drastic social and economic consequences. They note that, although positions are divided in the scientific community regarding the opportunities and limitations of changes resulting from digitization, the change caused by technology is having a profound effect on the education sector and the creation of policies/measures.

Wildemeersch and Jütte (2017) draw attention to the problems arising from the digital divide and the disparity in literacy rates between generations. They stress that policies to promote ICT are generally not multifaceted and that digitization has engendered other kinds of problems, particularly in the fields of privacy, data protection, freedom of expression, and a certain notion of "digital dictatorship" (for example, through the identification of people using biometric data). New issues of concern to society also extend to the provision of public and private knowledge, as well as to the field of education and learning opportunities, dependencies, and hierarchies.

Cardoso et al. (2021) also draw attention to the fact that due to the pandemic caused by the spread of the coronavirus on a global scale, the entire world population had to adapt and adjust their behaviors as quickly as possible to the evolution of the disease, particularly those at higher risk (for example, seniors)—"This vulnerability is due to the physiological process of ageing itself, which incites a decrease in the efficiency of the immune system, increasing the propensity for morbidity and mortality from infectious diseases" (p. 145). Thus, in order to combat the spread of the disease, the communication strategy focused on implementing a process of isolation on an unprecedented global scale with the objective of using mass communication and thus motivating individuals to adopt preventive behaviors and protect themselves and those around them.

Cardoso et al. (2021) emphasize that prevention of COVID-19 is largely dependent on actions structured by the official bodies which are responsible for the field of health, but it also depends on the behaviors that each individual adopts, or doesn't, in their social contexts. Seniors' low level of literacy is presented as one of the factors that makes them more vulnerable, given the diseases/conditions that come with aging. Therefore, these authors find that health literacy opens up "the possibility of providing safe self-care to the elderly not only to face the COVID-19 pandemic" (p. 150) but also in any situation in which decisions need to be made that concern their health.

Furthermore, Chau et al. (2010) carried out a survey to determine the health literacy education needs of aging populations in Hong Kong and concluded that, as in the West, also in Asia: (1) there are gaps in knowledge and misinterpretations regarding changes in health which are inherent to aging; (2) there is a tendency toward overconfidence in new technologies, as well as seniors being dependent "on others" as regards health matters; (3) ageism is a current form of discrimination; (4) seniors experience problems in consulting services; (5) there is a lack of knowledge regarding the type of home help available; and (6) there are unrealistic expectations about longevity. They also proved that the gap is more prevalent at older ages, although it has also been recorded among health sector professionals.

It can thus be stated that education and social awareness of seniors' problems should focus on these areas in order to increase health literacy rates and promote empowerment. Society can thus be better enabled to deal with the problems of aging and continue along the path that needs to be taken toward increased empowerment for senior populations.

5 Conclusions

The drive of health communication is the welfare of citizens. It aims to promote health, prevent disease, and raise awareness about new potential health issues, with the objective of carrying out effective interventions in communities and with key groups who are at greater risk, for example, seniors.

This area is revealed as having an extremely important position in the interaction, influence, and empowerment of individuals in matters related to their health, through information shared on social and digital media (Schiavo, 2014).

Seniors (aged 65 or over) have proved to be one of the most challenging groups for communication in health, as they are increasingly numerous due to the increase in average life expectancy; the mission is to satisfy the needs and demands of seniors and make it viable for the health system to support this effort (Berry, 2007). However, one of the biggest problems associated with aging has been shown to be discrimination, namely ageism. This term represents more than a mere negative attitude toward seniors; indeed, it is a profound reflection of the cultural values and institutional practices of our society (Marques, 2011).

In the Portuguese scenario, there is still a lack of specific state policies on the inclusion of seniors in the digital universe; however, senior universities have made an important initiative in combating the digital divide during pandemic times (Afonso et al., 2020).

Thus, taking into account that health literacy can lead to empowerment, it is known that this represents an important element in decision-making on health issues, because it helps individuals reflect and interact with their health systems (Espanha et al., 2015). In the context of the pandemic caused by COVID-19, the conclusion can be drawn, in the case of seniors, that the low level of literacy is one of the factors that makes them more vulnerable to the disease (Cardoso et al., 2021).

6 Active learning sections

(1) Suggested teaching assignments
  * Search for e-Health platforms in your national/regional context and analyze their influence and improvements brought to the lives of senior citizens
  * Verify if your local government has initiatives to promote the inclusion of senior citizens and the rise of digital literacy levels

(2) Recommended readings
  * Banksota, S., Healy, M. & Goldberg, E. (2020). 15 Smartphone apps to use while in isolation during the COVID-19 pandemic. *Western Journal of Emergency Medicine*, *0*(0), pp. 1–12. doi: https://doi.org/10.5811/westjem.2020.4.47372.
  * Fang, M. L. (2019). Exploring privilege in the digital divide: implications for theory, policy and practice. *The Gerontologist*, *59*(1), pp. 1–15
  * Espanha, R. (2020). Valor da Comunicação em Gestão Pública: Um Exemplo Aplicado à Área da Saúde Pública. In: FÉLIX, J. d'A. B. et al. *Comunicação Estratégica e Integrada*.
  * Marques, S. et al. (2020). Determinants of ageism against older adults: A systematic review. *International Journal of Environmental Research and Public Health*. doi:10.3390/ijerph17072560

(3) Case study

(1) App: Elsa (https://www.elsa.science/en/about-elsa/)

Description: Elsa is an application that helps users learn about rheumatoid arthritis (RA). Elsa is being codeveloped by people diagnosed with rheumatic conditions, researchers, healthcare providers, designers, and developers. There is also a community aspect that lets users interact and share their personal experiences in order to improve the content found in the app.

(2) App: Firstline—clinical decisions (https://firstline.org/)

Description: Firstline's community platform is purpose-built for healthcare professionals covering infectious diseases. Firstline is a customizable point-of-care tool/app for infectious diseases. It includes features such as:
  (i) Customizability for any hospital or institution;
  (ii) Antimicrobial stewardship resources;
  (iii) COVID-19 guidelines;

(iv) Infection prevention and control protocols;

(v) Antimicrobial formulary information;

(vi) Pathogen data including local antibiogram data;

(vii) Messaging system with push notifications;

(viii) Surveys and forms;

(ix) Being cloud based;

(x) Rapid updates;

(xi) Offline function.

(3) App: SNS 24 (https://www.sns.gov.pt/apps/sns24/) (Portugal)

Description: This app is built to gather Portuguese citizens' health information, namely, about vaccination, medical recipes, allergies, and medical exams. Users can also register specific health measurements (blood glucose, blood pressure, body mass index), teleconsultation, and medical recipe renewal.

(4) Titles for research essays

- "Medical apps in an ever-aging world: Health and digital literacy for senior citizens"
- "Digital Ageism—solutions, challenges, and perspectives to fight the digital abyss"
- "Seniors, E-health, and Empowerment: teaching digital habits to consolidate seniors' decision-making process"

(5) Recommended projects URL

(1) International Health Literacy Association (IHLA) (https://i-hla.org/)

(2) Health Literacy Survey (HLS) (https://gulbenkian.pt/publication/literacia-em-saude-em-portugal/)

References

Abroms, L. C., & Maibach, E. W. (2008). The effectiveness of mass communication to change public behavior. *Annual Review of Public Health*, *29*. http://arjournals.annualreviews.org/eprint/62aIzq67kfwRyWgv84VR/full/10.1146/annurev.publhealth.29.020907.090824.

Afonso, C., Fernandes, H., & Magalhães, C. (2020). IV – Inclusão digital do idoso: uma agenda para tempos de covid19 e para o futuro. In *V Conferência Científica Internacional de Projetos Educativos para Seniores – 2020*. Programme online at www.ripe50.org.

Araújo, I. S., & Cardoso, J. M. (2007). In Fiocruz (Ed.), *Comunicação e Saúde*.

Berger, P. L., & Luckmann, T. (2004). *A Construção Social da Realidade – Um livro sobre a sociologia do conhecimento*. Dinalivro.

Bernhardt, J. (2004). Communication at the core of effective public health. *American Journal of Public Health*, *94*, 2051–2053.

Berry, D. (2007). *Health communication – Theory and practice*. Open University Press.

Cardoso, R. S. S., et al. (2021). Letramento em saúde na pessoa idosa em tempos de pandemia e infodemia do COVID-19: um desafio mundial. *Enfermagem Gerontológica no Cuidado ao Idoso em Tempos da COVID-19*, *3*, 145–150. https://doi.org/10.51234/aben.21.e05.c21.

Chau, P. H., et al. (2010). Raising health literacy and promoting empowerment to meet the challenges of aging in Hong Kong. *Educational Gerontology*, *36*, 12–25. https://doi.org/10.1080/03601270902917547.

Diário da República. (2020). *Portugal Digital: Apresentação do Plano de Ação para a Transição Digital*. March https://www.portugal.gov.pt/gc22/portugal-digital/documento-de-suporte-a-apresentacao-realizada-a-5-de-marco-de-2020-pdf.aspx.

Dias, I. (2012). O uso das tecnologias digitais entre os seniores. Motivações e interesses. *Sociologia, Problemas e Práticas, 68,* 51–77.

Espanha, R. (2019). *A comunicação no âmbito da saúde pública. Aggregation exams.* ISCSP – U. Lisboa.

Espanha, R. (2020). Valor da Comunicação em Gestão Pública: Um Exemplo Aplicado à Área da Saúde Pública. In J. A. B. Félix, et al. (Eds.), *Comunicação Estratégica e Integrada.*

Espanha, et al. (2015). *Literacia em saúde em Portugal.* Fundação Calouste Gulbenkian.

Eysenbach, G. (2001). What is e-health? *Journal of Medical Internet Research, 3*(2).

Gamliel, T., & Gabay, N. (2014). Knowledge exchange, social interactions, and empowerment in an inter-generational technology program at school. *Educational Gerontology, 40,* 597–617.

Giddens, A. (2000). *As consequências da modernidade.* Celta.

Hill, R., et al. (2015). Older adults' experiences and perceptions of digital technology: (Dis)empowerment, wellbeing, and inclusion. *Computers in Human Behavior, 48,* 415–423.

Hou, S. (2012). Health literacy online: A guide to writing and designing easy-to-use health web sites. *Health Promotion Practice, 13*(5), 577–580.

Kreps, G. L. (2001). The evolution and advancement of health communication inquiry. In W. B. Gudykunst (Ed.), *24. Communication yearbook* (pp. 232–254).

Kreps, G. L., & Maibach, E. W. (2008). Transdisciplinary science: The nexus between communication and public health. *Journal of Communication, 58,* 732–748. https://doi.org/10.1111/j.1460-2466.2008.00411.x.

Lüders, M., & Brandtzæg, P. B. (2014). *My children tell me it's so simple: A mixed-methods approach to understand older non-users' perceptions of social networking sites.* New Media & Society.

Lugano, G., & Peltonen, P. (2012). Building intergenerational bridges between digital natives and digital immigrants: Attitudes, motivations and appreciation for old and new media. In E. Loos, L. Haddon, & E. Mante-Meijer (Eds.), *Generational use of new media* (pp. 151–170). Farnham: Ashgate Publishing, Ltd.

Marques, S. (2011). *Discriminação da Terceira Idade.* Fundação Francisco Manuel dos Santos.

Marques, S., et al. (2020). Determinants of ageism against older adults: A systematic review. *International Journal of Environmental Research and Public Health.* https://doi.org/10.3390/ijerph17072560.

Morris, A., Goodman, J., & Branding, H. (2007). Internet use and non-use: Views of older users. *Universal Access in the Information Society, 6*(43–57), 43–56.

Moscovici, S. (1978). In Zahar (Ed.), *A representação social da psicanálise.*

Moscovici, S. (2005). *Memória, imaginário e representações sociais.* Museu da República.

Noblin, A., & Rutherford, A. (2017). Impact of health literacy on senior citizen engagement in health care IT usage. *Gerontology & Geriatric Medicine, 3,* 1–8.

Parker, R., & Ratzan, S. (2019). Our future with democratization of health requires health literacy to succeed. *American Behavioral Scientist, 63*(7), 948–954. https://doi.org/10.1177/0002764218755834.

Rebelo, C. (2016). Exclusão digital sénior: histórias de vida, gerações e cultura geracional. *Revista Comunicando, 5*(1), 144–158.

Riva, G., et al. (2016). *Positive technology: The use of technology for improving and sustaining personal change.* IGI Global. https://doi.org/10.4018/978-1-4666-9986-1.ch001.

Rowlands, J. (1995). Empowerment examined. *Development in Practice, 5*(2), 101–108.

Sak, G., et al. (2017). Assessing the predictive power of psychological empowerment and health literacy for older patients' participation in health care: A cross-sectional population-based study. *BMC Geriatrics, 17*(59). https://doi.org/10.1186/s12877-017-0448-x.

Sales, M. B., et al. (2009). Learning by peers: An alternative learning model for digital inclusion of elderly people. In *9th IFIP TC 3 world conference on computers in education, Brazil – Presentation.*

Schiavo, R. (2014). *Health communication – From theory to practice.* Jossey-Bass.

Selwyn, N. (2012). Making sense of young people, education and digital technology: The role of sociological theory. *Oxford Review of Education, 38*(1), 81–96.

Silverstone, R. (1999). *Why study the media?.* Sage.

Wildemeersch, D., & Jütte, W. (2017). Editorial: Digital the new normal – Multiple challenges for the education and learning of adults. *European Journal for Research on the Education and Learning of Adults, 8*(1), 7–20.

CHAPTER 12

Health communication training of health professionals: From theory to practice

Nour Mheidly
Department of Communication, University of Illinois Chicago, Chicago, IL, United States

1 Introduction

Health communication is a rising field concerned with shaping and improving public health status. It employs communication strategies that aim at disease prevention and health promotion to enhance well-being and improve quality of life (Rimal & Lapinski, 2009). Holistically, health communication can also be viewed as a means for exchanging information related to the public health of individuals and communities (Pagano, 2016).

Today, health communication, as a discipline, is gaining more exposure in various fields, such as healthcare and academia. Many universities are introducing it as an autonomous major at both undergraduate and graduate levels. The COVID-19 pandemic was crucial in highlighting the role of adequate health communication as an essential strategy to mitigate the effects of misinformation, strengthen the position of healthcare professionals during disasters, and maintain a healthy relationship between the public and public health officials during lockdowns.

Technology is an essential pillar of health communication. Its rise in parallel with health communication has given this field more exposure and possibilities to grow and prosper. Technology allows healthcare providers and the public to connect digitally by sending, receiving, and discussing rising health concerns (Hu, 2015). In addition to its role in improving clinical practices, technology opens doors for virtual health/medical education (Hu, 2015) and contributes to more functional healthcare systems through advanced electronic medical record (EMR) systems (Pagano, 2016).

Overall, health communication contributes significantly to building "health literacy" among the public. Although the definition of the latter term varies between scholars, all the definitions agree that health literacy is fulfilled when an individual is educated about one's health. Freedman et al. (2009) viewed health literacy as the extent to which individuals and groups can comprehend, analyze, evaluate, and take actions and decisions about health issues that would benefit society. Similarly, Berkman, Sheridan,

Donahue, Halpern, and Crotty (2011) perceived health literacy as an essential component of successful health communication techniques that have a direct influence on public health. Developing skills in understanding and applying information about health issues have a substantial impact on health behaviors and health outcomes. Gaining these skills through proper health communication leads to better health literacy. It is recommended that professional schools such as medicine, nursing, dentistry, pharmacy, journalism, and social work include health literacy education in their curricula and emphasize its importance in improving human health (Nielsen-Bohlman, Panzer, & Kindig, 2004).

2 Main approach

This chapter will review the impact of health communication on the performance of healthcare professionals, such as doctors, nurses, practitioners, and medical students. The effect of proper health communication training is studied with respect to patient literacy rates and health needs. The role of health communication training in caring for vulnerable populations, including pregnant women and patients with cancer, is also discussed. The role of technology in health communication and healthcare settings is asserted. Finally, the significance of health communication training in promoting global public health and advocating for better health policy is explored.

3 Literature review

Health professionals are doctors, nurses, practitioners, or any other individuals working in the health sector. Health communication training is vital to keep the workflow going and to ensure a successful transition of information between the different components of the healthcare system and the healthcare system and the patients.

Health communication training programs are essential for enhancing healthcare outcomes and improving communication between health professionals and patients. Green, Gonzaga, Cohen, and Spagnoletti (2014) explored the efficacy of health communication curricula in improving the health literacy—skills and knowledge—of clinical residents. The training focused on didactic teaching material, videotaped individualized feedback from patients, and general practice with patients (Green et al., 2014). Results revealed that training increased the understanding of health literacy among residents and amplified their confidence in dealing with patients with low literacy rates (Green et al., 2014). Pagels et al. (2015) explored the effect of communication skills training curriculum for family medicine residents to enhance their relationship with patients with low health literacy rates. A group of 25 family medicine residents underwent health literacy and communication training that was backed up with didactic materials and resources (Pagels et al., 2015). After comparing and analyzing the pre- and post-knowledge of residents, results showed that residents' health literacy knowledge increased as well as their self-confidence while dealing with patients with low health literacy (Pagels et al., 2015). Selling et al. (2021) explored the effects of implementing relationship-centered care (RCC) communication curriculum training and coaching for pediatric residents. A group of 77 residents participated in a 4-h RCC training and filled up 2 surveys, one directly after the training and the other after 6 months. After 6 months, residents showed a significant increase in the use of RCC, as they mentioned that their perspective on RCC training has changed and shifted to a positive one (Selling et al., 2021). Combining RCC communication training with coaching further improves patient communication skills and helps residents adapt their knowledge in different settings (Selling et al., 2021).

Communication training enhances the communication skills of health professionals when dealing with patients that need sensitive care. Gysels, Richardson, and Higginson (2004) studied the effect of communication training on health professionals that take care of patients with cancer. The study evaluated the available communication training literature in the field of cancer. Out of 47 studies, 16 papers described 13 interventions, 4 of which were explored using randomized controlled trials (Gysels et al., 2004). Results revealed that most interventions improved basic communication between health professionals and patients with cancer (Gysels et al., 2004). Dieterich and Demirci (2020) studied health communication practices that healthcare professionals adhere to when dealing with overweight pregnant women. They explored research from 2008 until 2018 on counseling and communication methods of health professionals with obese pregnant women. Results showed that health professionals are hesitant to communicate with pregnant women about their extra weight, as they are uncomfortable and lack self-confidence when discussing these topics. As a result, effective centered health communication training is helpful in improving the communication skills of health professionals while dealing with obese pregnant women. Furthermore, this training will help obese pregnant women feel less stigmatized about their weight while focusing on improving their lifestyle regimen.

The current status of sexual health communication is suboptimal due to the stigma, shame, and sensitivity that healthcare professionals and patients face when discussing it. Engelen, Knoll, Rabsztyn, Maas-van Schaaijk, and van Gaal (2020) conducted a systematic review to explore the elements used in sexual communication between health professionals and patients with chronic conditions. Only 15 studies were collected from 2000 until 2018. The findings of the study showed that there are four major determinants for sexual communication, namely knowledge, attitude, self-efficacy, and beliefs, which can be utilized to enhance and trigger sexual health communication in healthcare settings (Engelen et al., 2020). Furthermore, the findings suggested the need for more research investment in the field and recommended integrating sexual education in all hospital curricula (Engelen et al., 2020).

Unpaid and untrained informal caregivers compose a major part of the healthcare system. Terui, Goldsmith, Huang, and Williams (2020) addressed the importance of health communication training for patients and informal caregivers by offering caregivers the opportunity to participate in three training workshops. These workshops stimulated topics and conversations about their health challenges and helped them express their concerns, understand medical conditions, and seek health support from the right parties (Terui et al., 2020).

Health communication training for health professionals is essential in promoting vaccinations among adolescents. Dempsey et al. (2018) studied the effects of healthcare professional communication training in increasing intervention and awareness of Papillomavirus vaccination (HPV). After performing a randomized controlled trial, healthcare professionals reported that communication training was an efficient intervention to enhance the intake of HPV vaccine among adolescents (Dempsey et al., 2018).

Training healthcare providers to improve their communication skills has a direct effect on the level of satisfaction of the patients. Pedersen, Brennan, Nance, and Rosenbaum (2021) explored the effect of communication skills training (CST) for healthcare professionals on the satisfaction score of patients. Physicians, physician assistants, and nurses were the participants that underwent training in addition to in-person coaching sessions (Pedersen et al., 2021). Satisfaction rates of patients were measured before and after the study. Results showed that the satisfaction rates of patients regarding healthcare providers that underwent CST were higher (Pedersen et al., 2021). In addition, participants reported being more self-confident in translating the skills they learned in the workshops into their clinical practice (Pedersen et al., 2021). Wolderslund, Kofoed, and Ammentorp (2021) explored using person-centered CST on a large network of healthcare professionals in Denmark. Healthcare professionals responded to a pre- and post-training questionnaire to measure their self-efficacy progress. Results showed that implementing CST significantly enhanced the health professional's self-efficacy and ensured that implementing a larger scale evidence-based training is as effective as small, controlled groups (Wolderslund et al., 2021).

Nurses are considered an essential part of the healthcare system. Providing them with health communication training helps them stay motivated, elevates their productivity, and boosts their self-confidence when dealing with patients. Wittenberg, Goldsmith, Buller, Ragan, and Ferrell (2019) conducted a cross-sectional study to explore the status of oncology nurses' communication practices with patients. Findings showed that the communication status and role of the nurse in the eyes of patients were unclear, highlighting the need for training programs that can improve communication between the two parties involved (Wittenberg et al., 2019). Nantsupawat et al. conducted a cross-sectional study to examine the nurses' understanding and knowledge of health literacy and communication methods and the barriers they face when applying for health literacy programs in healthcare settings. A total of 1697 nurses in community hospitals from

Thailand completed the study. Results revealed that 55.3% of the nurses were familiar with the term "health literacy," 51.7% had a moderate knowledge of health literacy, and 90.6% hadn't undergone any type of formal training that helps them communicate with patients with low literacy rates. The barriers to implementing health communication and literacy programs included a lack of health literacy specialists, training and assessment tools, educational resources, and a scarcity of time (Nantsupawat et al., 2020).

Research on health training for health professionals is important in exploring the status of communication training, finding the gaps, and proposing recommendations to enhance the competencies of health care professionals. Singh, Kennedy, and Stupans (2020) explored the competencies of health professionals that are responsible for taking care of patients with chronic health conditions. Results showed that there are nine main competencies that health professionals must meet that enhance health outcomes: effective communication with patients, activity coordination, leadership and team skills, confidence, accountability, understanding evidence-based knowledge, respecting colleagues and patients with different backgrounds, identifying gaps, and working on areas that need development (Singh et al., 2020).

4 Discussion
4.1 Significance of health communication training

The use of health communication can be traced back to the beginning of humanity (Finnegan, 1989). Communication between people about injuries, diseases, food, and threats surrounding them was an application of health communication practices (Kreps, Bonaguro, & Query, 2003). In the 20th century, members of the International Communication Association were the first to adopt the "health communication" term. Later, the uprising of public health policy and nongovernmental organizations (NGOs) and other organizations expanded the role of health communication, making it one of the essential fields that help people survive the health threats of the 21st century (Atkin & Silk, 2009). Dealing with existing global health risks such as tobacco and substance use, overweight and obesity, pollution, sanitization, unsafe sex, and the diffusion of infectious diseases necessitated raising awareness through health communication campaigns to educate and spread health and wellbeing-related topics to the public (CDC, 2019; WHO, 2009) (Fig. 1). Although the field is on the rise and is gaining more exposure than before, many countries and regions are still lagging in terms of health communication literacy and research. In low-income countries, health communication research comprised 0.27% of the world's health communication research (Mheidly & Fares, 2020a). The Arab world's health communication research constituted only 1% of the world's health communication research (Mheidly & Fares, 2020b). As such, barriers that stand in the way of advancing health communication education globally must be recognized and addressed to advance the field and healthcare practices forward.

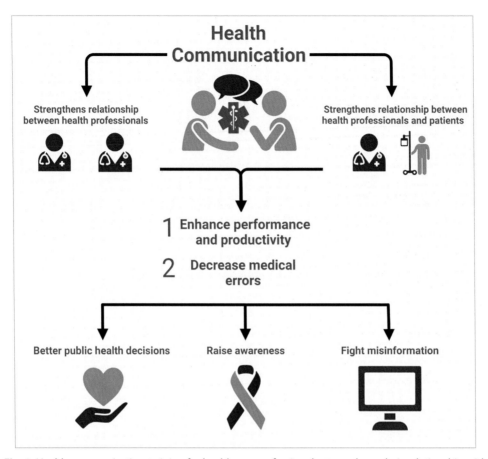

Fig. 1 Health communication training for healthcare professionals strengthens their relationship with the patients and between health professionals themselves, improves their performance and productivity, and decreases medical errors. This, in turn, leads to better public health decisions, more awareness, and less misinformation.

4.2 Integration of technology in health communication training

The technological advancements that took place in the 21st century improved health communication and facilitated the dissemination of important health topics to the public. Technology played a major role in improving the communication practices between individuals, making it easier to distantly connect with anyone at any anytime. It further strengthened the educational sector and opened new prospects for people to learn and gain knowledge in many disciplines. For example, during the COVID-19 pandemic, social media platforms were the most used tools to spread awareness, information, and instructions on the latest COVID-19 news and recommendations. Twitter and

Facebook, specifically, were effective platforms for communicating with the public about healthcare topics (Ades, 2021).

Advancements in technology allowed online/distance health education. During the pandemic, classes, conferences, workshops, and meetings were all held virtually (Danchikov, Prodanova, Kovalenko, & Bondarenko, 2021). This shift to online learning permits healthcare professionals to learn health communication skills synchronously or asynchronously in the comfort of their own settings. Nevertheless, health professionals need to be trained and well-prepared for the online shift. Sethi, Sethi, Ali, and Aamir (2020) explain that Pakistani health professionals and faculty members, for instance, were untrained for online learning during the COVID-19 pandemic, recalling that some were even technophobic. Online learning necessitates curriculum reconstruction for online activities to enhance and develop health practices and clinical outcomes in society (Seymour-Walsh, Bell, Weber, & Smith, 2020).

Technology has also given the opportunity for health professionals to train on simulators to master communication skills and address health inequities. Simulators can be used for practicing and task-training on standardized patient scenarios (Galloway, 2009). Billon et al. (2016) explored the effectiveness of simulation training courses for health professionals to support patients with intellectual disabilities. Results were measured before and after the simulation course, by which time significant improvements were observed in participants' communication skills, attitudes, and knowledge of people with disability (Billon et al., 2016). Therefore, including simulation in health communication curricula of health professionals can increase their knowledge and health communication practices with their peers and patients.

Technology improved healthcare professionals' knowledge and understanding of their own health and that of their patients. Pai and Alathur (2019) assessed the effectiveness of mobile phone technology in health and wellness delivery among a sample of health professionals, medical students, technical students, and working staff in India. Results revealed that most females used their mobile phones to track their menstrual cycle and check information about diseases (Pai & Alathur, 2019). Nevertheless, results showed that there was a lack of use of mobile phones for health awareness and education (Pai & Alathur, 2019). As such, training healthcare professionals, early on, on the use of their mobile devices to improve their practices is vital.

Today, the focus is on visuals. People are more attracted to graphics, especially in healthcare educational settings, where an image can be a successful tool to deliver health messages in a fast, straightforward manner (Houts, Doak, Doak, & Loscalzo, 2006). Software engineering and technological advancement eased the design and creation of such visual material. As such, it is vital for health professionals to be trained to plan, design, and create pictures and infographics to help patients with low literacy rates be well-informed on the latest public health emerging topics.

4.3 Role of health communication in global public health and policy

Governments and NGOs are important communicators of health information. They promote awareness, educate, and enhance the propagation of global health issues to the public. They play an essential role in developing vaccines, disease prevention strategies, and providing health information to influence public opinion (Klugman, 2000). NGOs can further help translate research to action by promoting global health awareness to reduce the disease burden and maintain an acceptable level of health literacy among the public (Delisle, Roberts, Munro, Jones, & Gyorkos, 2005). Consequently, it is important for healthcare professionals to be familiar with how governmental organizations and NGOs work to address public health needs. Implementing training programs and educational curricula led by government and NGO health officers can help healthcare professionals improve their communication with surrounding communities and vulnerable populations.

5 Conclusion and recommendations

Health communication training for healthcare professionals is essential for many reasons. It strengthens their relationship with the patients and builds more trust between them, strengthens the relationship between health professionals themselves, improves their performance and productivity, decreases medical errors, influences better public health decisions, helps raise awareness, especially in underperceived communities, and helps fights misinformation. The rise of technological advancements has enhanced the field of health communication and increased its impact, especially through social media platforms and multimedia outlets. It has also allowed online education and simulation training of health professionals to improve their health communication skills in the work setting. The leading role played by governments and NGOs in drafting health communication strategies at the national and global level necessitates the exposure of health professionals to such practices and their enrollment in training programs led by such organizations to improve their understanding and impact of health communication measures.

6 Active learning sections

1. **Suggested teaching assignments:** Search the internet for academic publications and/or governmental reports on the status of health communication curricula in healthcare settings in your own country. Then, draft a one-page summary report to answer the following:
   - What is the status of health communication curricula in your country?
   - What problems/challenges are faced in health communication training in your country?
   - Suggest solutions/recommendations to improve health communication in your country.

2. **Recommended readings:** Network of the National Library of Medicine [NNLM]. (2020, October 21). Effective Health Communication and Health Literacy: Understanding the Connection [Video]. YouTube. https://www.youtube.com/watch?v=82DqnjphXGY&ab_channel=NetworkoftheNationalLibraryofMedicine%5BNNLM%5D

   At the end of the video, students will be able to:
   * Define health literacy
   * Describe universal precautions for health literacy
   * Name three components of clear health communication
   * Identify three online resources you can use as tools to promote health literacy

3. **Case study:** Explore this case and answer the following questions: Altman, L., (1995). Big doses of chemotherapy drug killed patient, hurt 2d. *The New York Times*, 24.
   * Identify the error in communication
   * Identify the consequences of this miscommunication
   * As a health communication advocate, how would you have avoided the miscommunication in the above case?

4. **Relevant research essays titles:**

   Book: Schiavo, R., (2013). *Health communication: From theory to practice* (Vol. 217). John Wiley & Sons.

   Article: Gysels, M., Richardson, A., & Higginson, I.J., (2004). Communication training for health professionals who care for patients with cancer: a systematic review of effectiveness. *Supportive Care in Cancer, 12*(10), 692–700.

5. **Recommended Projects URL:**
   * Health Literacy Solution Center: https://www.healthliteracysolutions.org/learning-lab/overview
   * Institute for Healthcare Communication: https://healthcarecomm.org/

References

Ades, A. S. (2021). The effective of health communication about the awareness of COVID-19 through social media. *Social Medicine, 13*(3), 118–126.

Atkin, C., & Silk, K. (2009). Health communication. In D. W. Stacks, & M. B. Salwen (Eds.), *Communication theory and methodology* (2nd ed., pp. 489–503). Routledge.

Berkman, N. D., Sheridan, S. L., Donahue, K. E., Halpern, D. J., & Crotty, K. (2011). Low health literacy and health outcomes: An updated systematic review. *Annals of Internal Medicine, 155*, 97–107.

Billon, G., Attoe, C., Marshall-Tate, K., Riches, S., Wheildon, J., & Cross, S. (2016). Simulation training to support healthcare professionals to meet the health needs of people with intellectual disabilities. *Advances in Mental Health and Intellectual Disabilities, 10*(5), 284–292.

CDC. (2019). *Health communication basics: centers for disease control and prevention.* Available: https://www.cdc.gov/healthcommunication/healthbasics/index.html.

Danchikov, E. A., Prodanova, N. A., Kovalenko, Y. N., & Bondarenko, T. G. (2021). Using different approaches to organizing distance learning during the COVID-19 pandemic: Opportunities and disadvantages. *Linguistics and culture review, 5*(S1), 587–595.

Delisle, H., Roberts, J. H., Munro, M., Jones, L., & Gyorkos, T. W. (2005). The role of NGOs in global health research for development. *Health Research Policy and Systems, 3*(1), 1–21.

Dempsey, A. F., Pyrznawoski, J., Lockhart, S., Barnard, J., Campagna, E. J., Garrett, K., et al. (2018). Effect of a health care professional communication training intervention on adolescent human papillomavirus vaccination: A cluster randomized clinical trial. *JAMA Pediatrics, 172*(5), e180016.

Dieterich, R., & Demirci, J. (2020). Communication practices of healthcare professionals when caring for overweight/obese pregnant women: A scoping review. *Patient Education and Counseling, 103*(10), 1902–1912.

Engelen, M. M., Knoll, J. L., Rabsztyn, P. R., Maas-van Schaaijk, N. M., & van Gaal, B. G. (2020). Sexual health communication between healthcare professionals and adolescents with chronic conditions in Western countries: An integrative review. *Sexuality and Disability, 38*(2), 191–216.

Finnegan, J. (1989). Health and communication: Medical and public health influences on the research agenda. In E. Ray, & L. Donohew (Eds.), *Communication and health: Systems and applications* (pp. 9–24). Hillsdale, NJ: Erlbaum.

Freedman, D. A., Bess, K. D., Tucker, H. A., Boyd, D. L., Tuchman, A. M., & Wallston, K. A. (2009). Public healthliteracy defined. *American Journal of Preventive Medicine, 36*(5), 446–451.

Galloway, S. (2009). Simulation techniques to bridge the gap between novice and competent healthcare professionals. *Online Journal of Issues in Nursing, 14*(2).

Green, J. A., Gonzaga, A. M., Cohen, E. D., & Spagnoletti, C. L. (2014). Addressing health literacy through clear health communication: A training program for internal medicine residents. *Patient Education and Counseling, 95*(1), 76–82.

Gysels, M., Richardson, A., & Higginson, I. J. (2004). Communication training for health professionals who care for patients with cancer: A systematic review of effectiveness. *Supportive Care in Cancer, 12*(10), 692–700.

Houts, P. S., Doak, C. C., Doak, L. G., & Loscalzo, M. J. (2006). The role of pictures in improving health communication: A review of research on attention, comprehension, recall, and adherence. *Patient Education and Counseling, 61*(2), 173–190.

Hu, Y. (2015). Health communication research in the digital age: A systematic review. *Journal of Communication in Healthcare, 8*(4), 260–288.

Klugman, B. (2000). The role of NGOs as agents for change. *Development Dialogue,* (1/2), 95–120.

Kreps, G. L., Bonaguro, E. W., & Query, J. L. (2003). The history and development of the field of health communication. *Russian Journal of Communication, 10,* 12–20.

Mheidly, N., & Fares, J. (2020a). Health communication in low-income countries: A 60-year bibliometric and thematic analysis. *Journal of Education and Health Promotion, 9*(2020), 163.

Mheidly, N., & Fares, J. (2020b). Health communication research in the Arab world: A bibliometric analysis. *Integrated Healthcare Journal, 2*(1), e000011.

Nantsupawat, A., Wichaikhum, O. A., Abhicharttibutra, K., Kunaviktikul, W., Nurumal, M. S. B., & Poghosyan, L. (2020). Nurses' knowledge of health literacy, communication techniques, and barriers to the implementation of health literacy programs: A cross-sectional study. *Nursing & Health Sciences, 22*(3), 577–585.

Nielsen-Bohlman, L., Panzer, A. M., & Kindig, D. A. (Eds.). (2004). *Health literacy: A prescription to end confusion.* Washington, DC: National Academies Press.

Pagano, M. (2016). *Health communication for health care professionals: An applied approach.* New York: Springer.

Pagels, P., Kindratt, T., Arnold, D., Brandt, J., Woodfin, G., & Gimpel, N. (2015). Training family medicine residents in effective communication skills while utilizing promotoras as standardized patients in OSCEs: A health literacy curriculum. *International Journal of Family Medicine, 2015,* 129187. https://doi.org/10.1155/2015/129187.

Pai, R. R., & Alathur, S. (2019). Assessing awareness and use of mobile phone technology for health and wellness: Insights from India. *Health Policy and Technology, 8*(3), 221–227.

Pedersen, K., Brennan, T. M., Nance, A. D., & Rosenbaum, M. E. (2021). Individualized coaching in health system-wide provider communication training. *Patient Education and Counseling, 104*(10), 2400–2405.

Rimal, R. N., & Lapinski, M. K. (2009). Why health communication is important in public health. *Bulletin of the World Health Organization, 87,* 247.

Selling, S. K., Kirkey, D., Goyal, T., Singh, A., Gold, C. A., Hilgenberg, S. L., et al. (2021). Impact of a relationship-centered care communication curriculum on pediatric residents' practice, perspectives, and opportunities to develop expertise. *Patient Education and Counseling, 105*(5), 1290–1297.

Sethi, B. A., Sethi, A., Ali, S., & Aamir, H. S. (2020). Impact of coronavirus disease (COVID-19) pandemic on health professionals. *Pakistan Journal of Medical Sciences*, *36*(COVID19-S4), S6.

Seymour-Walsh, A. E., Bell, A., Weber, A., & Smith, T. (2020). Adapting to a new reality: COVID-19 coronavirus and online education in the health professions. *Rural and Remote Health*, *20*(2), 6000.

Singh, H. K., Kennedy, G. A., & Stupans, I. (2020). Competencies and training of health professionals engaged in health coaching: A systematic review. *Chronic Illness*, *18*(1), 58–85.

Terui, S., Goldsmith, J. V., Huang, J., & Williams, J. (2020). Health literacy and health communication training for underserved patients and informal family caregivers. *Journal of Health Care for the Poor and Underserved*, *31*(2), 635–645.

Wittenberg, E., Goldsmith, J., Buller, H., Ragan, S. L., & Ferrell, B. (2019). Communication training: Needs among oncology nurses across the cancer continuum. *Clinical Journal of Oncology Nursing*, *23*(1).

Wolderslund, M., Kofoed, P. E., & Ammentorp, J. (2021). The effectiveness of a person-centred communication skills training programme for the health care professionals of a large hospital in Denmark. *Patient Education and Counseling*, *104*(6), 1423–1430.

World Health Organization. (2009). *Global health risks: Mortality and burden of disease attributable to selected major risks*. World Health Organization.

CHAPTER 13

Nursing advanced training: Person-centeredness and technology for innovative pedagogy

Tânia Manuel Moço Morgado[a,b,c,d,e], Filipa Isabel Quaresma Santos Ventura[b,c], Rosa Carla Gomes Silva[e,f], Hugo Leiria Neves[b,c,f], Joana Sofia Dias Pereira Sousa[d,g], and Pedro Miguel de Almeida Melo[h]

[a]Centro Hospitalar e Universitário de Coimbra, Coimbra, Portugal
[b]Health Sciences Research Unit: Nursing (UICISA: E), Nursing School of Coimbra, Coimbra, Portugal
[c]Nursing School of Coimbra, Coimbra, Portugal
[d]School of Health Sciences, Polytechnic of Leiria, Leiria, Portugal
[e]Center for Health Technology and Services Research (CINTESIS), Porto Nursing School (ESEP), Porto, Portugal
[f]Portugal Centre for Evidence Based Practice: A Joanna Briggs Institute Centre of Excellence, Coimbra, Portugal
[g]Center for Innovative Care and Health Technology - CiTechCare, Leiria, Portugal
[h]Universidade Católica Portuguesa, Centre for Interdisciplinary Research in Health/Institute of Health Sciences, Porto, Portugal

1 Introduction

People around the world have the right to high-quality, safe, and affordable healthcare. The best health outcomes, including adherence to treatment and healthy lifestyles, can be achieved when healthcare professionals, and particularly nurses, focus on the real needs of citizens or patients. In real-world practice, planning and implementing healthcare aimed at the experienced needs of the citizen or patient is not possible without including the family/caregiver and the community, where each person belongs. Throughout history, we can see the continuing transformation of the nursing discipline and profession evolving to require more advanced nursing practice along with the health, societal, and person-centered care challenges (International Council of Nurses (ICN), 2020).

Advanced nursing practice is underpinned on discipline-specific theoretical knowledge, which draws on philosophical perspectives and ontological, epistemological, and methodological frameworks, and is driven by an ethical approach toward humans and the world they inhabit (Parse, 2016; Parse et al., 2000). Advanced Practice Nursing, according to the ICN (2020), is considered as advanced nursing interventions that influence clinical outcomes for individuals, families, and diverse populations, and is based on education and training on central criteria and core competencies for practice (Canadian Nurses Association, 2019; Hamric & Tracy, 2019), critical thinking and understanding of the theoretical background (Parker & Hill, 2017). An Advanced Practice Nurse is a generalist or specialist nurse who has acquired, the expert knowledge base, complex decision-making skills, and clinical competencies for Advanced Practice Nursing, through post-graduate education (i.e., minimum of a master's degree) (ICN, 2020).

In the various areas of assistance in the nursing discipline and profession, the advanced training of nurses has always been needed and will continue to demand the best evidence, as it adds value to nursing curricula.

This chapter aims to clarify the importance of a person-centered curriculum and pedagogy based on the best evidence on the advanced training of nurses, starting from the needs of citizens, family/caregivers, and the community, and making the transition from person-centered nursing to a person-centered curriculum and evidence-based learning and research.

2 Supporting the person and the caregiver: Challenges to be addressed in Advanced Practice Nursing for sustainable healthcare systems

Person-centered practice is an essential cornerstone for sustainable healthcare systems (Lloyd et al., 2020) and a core competence of nurses (Bartz, 2010). Person-centered practice is an ethical and moral obligation and demands seeing the person and the family as unique human beings with both vulnerabilities and resources. Seeing the person before the disease, i.e., taking into account the experience and impact of the disease in the person's and family's daily life, is crucial to adequately meet the person's needs according to his/her preferences and values (Britten et al., 2020).

The practice of compassionate and person-oriented care might be considered a distinctive dimension of care rather than service-oriented and efficiency-driven (Phelan et al., 2017), which demands learning and training, as any other competence must be acquired, developed, and maintained (Phelan et al., 2020).

The best health outcomes, including adherence to treatment and healthy lifestyles, can be achieved when nurses focus on the real needs of citizens or persons (Chesak et al., 2021). In practice, it is impossible to plan and implement healthcare focused on the real needs of citizens or persons without including the family and the community where each one belongs (Chesak et al., 2021; Clemmensen et al., 2021). Supporting the caregivers is one of the challenges that need to be contemplated in the competency profile of Advanced Practice Nursing, if aiming to tackle the forthcoming demands of healthcare systems.

The self-management of health and illness is a challenging process for the citizen, given that it is not a subject domain that one would learn at school. Citizens need to be supported in learning to self-manage their health/illness processes. Therefore, the health professionals' deontological and moral obligation is to become the most competent professionals to have the tools and strategies to enable each person's self-management processes. Should a person not be capable of self-managing their health or life due to physical, mental, or cognitive incapacity, the nurse will similarly support the caregiver to exercise their role (Chesak et al., 2021; Clemmensen et al., 2021; Langins & Borgermans, 2015; Peterson, 2003; Sabo & Chin, 2021).

In the United States, more than 41 million family caregivers are estimated to spend an average of 20 hours a week taking care of their loved ones (Ng & Indran, 2021). The caregiver needs to be an active agent in society and motivated to make choices concerning complex diagnostic and treatment processes, among other challenges that the role entails. However, to have motivated and competent caregivers, healthcare professionals need to be highly prepared to achieve these outcomes (Langins & Borgermans, 2015; Sabo & Chin, 2021; Shin et al., 2018).

Uninformed caregivers who cannot play their role suffer from frustration, anguish, isolation and fatigue, depression and health problems (Anagnostou, 2021; Clemmensen et al., 2021; Ng & Indran, 2021). Healthcare professionals and informal caregivers need to work as a team aiming at the single goal of caring for the person. How can nurses be trained to support the caregiver? What is the expected competencies profile for nurses that support the caregiver in a person-centered manner and assist them in their role and activities? What innovative tools are at nurses' disposal to facilitate caregiver training and support?

Nurses' competencies must be developed in order to boost support for caregivers. The World Health Organization (WHO) (2006) describes competencies as tasks that different types and levels of health professionals can perform due to their training process. The skills and experiences reflect the educational background regarding the nontechnical qualities (e.g., compassion and motivation) and techniques necessary for the effective delivery of health services of each professional core (WHO, 2006). For this reason, these same skills embody the identity of each profession.

In operational terms, the ability to perform efficiently the activities expected of a nurse can be explained by levels of depth (i.e., identifiable—implementable—teachable), defining the basic set of information (knowledge), skills (knowing how to do), attitudes (knowing how to be), and experience necessary to reach a certain level of capacity and performance in the interaction established with the caregiver, regardless of their educational, cultural, or experiential profile for the performance of their role. "Thus, a competence indicates being able to do something well—measured against a standard—especially a capacity acquired through experience or training", p. 40 (Pan American Health Organization and World Health Organization, 2003).

Considering nursing as an impactful healthcare workforce, which reflects on transversal skills taught and developed at universities, it is imperative to develop an intervention targeting the caregiver (Anagnostou, 2021; Chesak et al., 2021).

The process of adapting the skills of the nursing discipline to the caregiver's needs implies ensuring that nurses have theoretical knowledge and practical skills to work efficiently and effectively with the caregiver. At the same time, it also implies that the healthcare workforce can apply this knowledge and skills in practice to consolidate their competencies, which will be constantly changing as the caregiver profiles are broad and rapidly evolving. So far, the focus on skills consolidation has been limited to the initial training of each healthcare professional individually during the training process at colleges/universities. That is, competencies are seen in isolation within each discipline. However, a broader and more comprehensive view, focusing beyond each discipline (e.g., psychology or nursing) and and adapted to each patient/caregiver, is necessary for sustainable healthcare systems (Chesak et al., 2021; Sabo & Chin, 2021).

Competence means adequacy and attitude, which in turn translates into being adequate and being able to, i.e., the "condition of being able" or "skill" and having a "specific range of skill, knowledge or skill" (Langins & Borgermans, 2015). Therefore, competencies serve to inform the standards by which the performance of each healthcare professional can be certified. A consistent, explicit, and common performance pattern for each particular discipline is necessary for this line of thought.

Furthermore, nursing curricula must be sufficiently open to a cross-cutting pattern of competencies common to the healthcare workforce.

In summary, the standard of competencies on caregiver support in interdisciplinary action in healthcare is necessary since the world is increasingly globalized and dynamic. Equally important to standardize common skills is the recognition of their flexibility and specificity within each profession. If an unclear definition of competencies can compromise health outcomes, a strict definition can hinder person-centered care and practice and limit the focus on persons' real needs. The academy, healthcare professions and particularly nurses need to reflect on yet another major issue: the changing healthcare needs of the population. That is, the skills of the various healthcare professionals need to be adapted to the society's change. Literacy levels have changed, health/illness processes are different as people live longer, and technology and innovation impose transformation of workflows. Such societal changes require adjustments in nurses' general and specialized skills (Langins & Borgermans, 2015; Ng & Indran, 2021; Parmar et al., 2021; Peterson, 2003; Sabo & Chin, 2021).

Nurses are uniquely positioned to work closely with caregivers within the healthcare workforce to support them in providing informed, safe, and sound care (Chesak et al., 2021). Accordingly, caregivers and the wider society demand nurses who need to be increasingly flexible to work in an interdisciplinary team, increasingly focused on the real needs of the caregiver, and capable of mobilizing and transferring the best evidence to practice. At the same time, these nurses must be attentive to innovative tools within the emerging paradigm of digital health and big data, including artificial intelligence, thus acting as a pivot who aims to meet real caregiver needs (Ronquillo et al., 2021; Shin, 2018).

Accordingly, advanced training in nursing using the best technologies is necessary for this area, where all actors play a crucial role. These actors are academia, students, healthcare systems, and caregivers, all walking together in this care partnership (i.e., healthcare professional-caregiver). The research processes in this domain must explore innovative tools for an increasingly closer intervention to the real needs. Simultaneously, academics and professional regulators must seek the best knowledge and reach a consensus on the skills that need to be developed transversely across nursing education and processes for joint support to the caregiver.

While carrying out their activities, nurses face major challenges in the caregivers' domain. Therefore, a reference frame of competencies for nurses is articulated and aligned between the various healthcare professionals, which seems to be an essential strategy for multidisciplinary, technology-mediated, and person-centered healthcare.

3 Person-centered curriculum

Preparing students to meet the complex needs of the person and family in a rapidly changing society is crucial to ensure individual-relevant, high-quality, evidence-informed, and safe healthcare (O'Donnell, McCormack, McCance, & McIlfatrick, 2020). Therefore, introducing person-centeredness in the nursing curriculum is essential to establishing lifelong professional development toward advanced practice (McCormack & Dewing, 2019). In its essence, a person-centered curriculum in nursing will seek to prepare students to exercise compassionate care and "care with others" rather than "care for others" and empower them to be transformative agents of practice in healthful environments (Dickson et al., 2020).

One major critique in nursing curriculum development is the lack of explicitness of conceptual, theoretical frameworks, and person-centered principles underpinning all stages of educational programs (i.e., content, scope, assessment, teaching and learning strategies) (O'Donnell et al., 2020). Toward enhancing explicitness and keeping in mind the graduate nurse that the program aims to scaffold, curriculum developers need to attend to (a) dimensions of person-centeredness and (b) person-centered care philosophy, in addition to (c) theoretical approaches to learning and teaching (Middleton & Moroney, 2019).

The International Community of Practice for Person-centered Practice (PcP-ICoP) (McCormack & Dewing, 2019) is one joint effort to embed person-centered principles into healthcare curriculum development. The PcP-ICoP came to a consensus of 10 core dimensions of person-centeredness that curriculum developers should attend to enhance consistency of discourse and ensure a systematic and transparent approach to the development, implementation, and evaluation of person-centered educational programs (McCormack & Dewing, 2019). By following these dimensions, curriculum developers will enhance the explicitness of person-centered conceptual elements and their interconnection. This may ensure that personalism permeates the curriculum coherently.

The self-awareness of values and beliefs, respect for diversity, active engagement in learning, and authentic relationships with others are some of the elements transpiring from the person-centeredness dimensions that a person-centered nursing curriculum should aim to foster.

The Person-centered Practice Framework (PcPF) (McCormack & McCance, 2017) might serve as the philosophical lens of person-centered care (Middleton & Moroney, 2019). The PcPF has its foundations in values of mutual respect for persons and the right to self-determination, which are fostered by empowering organizational cultures, where quality improvement of practice is continuously addressed and welcome (McCormack & McCance, 2017). The PcPF maps the person-centeredness elements at the levels of pre-requisites, care environment, and processes that lead to healthful cultures of human flourishing for all persons (i.e., patients, healthcare professionals, students, teachers). These dimensions are further considered at the macro level, where supportive elements of person-centered systems and political and strategic frameworks are identified (McCormack & McCance, 2017).

Finally, an educational philosophy aligned with the person-centeredness dimensions and person-centered practice philosophy is essential to bring together strategies of teaching and learning that facilitate the development of students' autonomy in a shared meaning-making environment toward becoming transformative graduates (Dickson et al., 2020). Constructivist learning theories might serve that purpose while focusing on the importance of continuous shared knowledge development in the world (Middleton & Moroney, 2019).

Altogether, a person-centered curriculum should foster competent, committed healthcare professionals while being attentive to the supportive local environment and *meso* and macro contexts (Dickson et al., 2020). Attending the relational dimension, a person-centered approach to curriculum design should encourage connectivity with the person's values, others, and context. In a coconstructionist fashion, both students and teachers/mentors are learners in a spiral journey toward achieving the purpose of scaffolding person-centered healthcare graduates. Along the learning journey, the forthcoming intuitive and reflexive healthcare professionals, together with their facilitators, will meaningfully engage with the many domains of the PcPF toward enhancing the person-centeredness of their practice (Dickson et al., 2020).

4 The value of research in nursing curricula and pedagogy based on the best evidence

Along with the increasing complexity of healthcare and facing a social value growth, the nursing profession had to define its disciplinary body of evidence within the multidisciplinary team early on. Research's value within the nursing discipline, profession, and the wider healthcare system is indisputable, with regional and global impacts (Coster, Watkins, & Norman, 2018; Yanbing et al., 2021).

Nursing has a long history of developing clinical research, particularly instigated by the academic processes of nurses, such as research carried out during undergraduate, master, and doctoral degrees; within the disciplinary area; or in related areas. Nevertheless, nursing research has become more robust and competitive, with more systematized processes especially fostered by the emergence of research units hosted in nursing education facilities (e.g., universities or higher education schools) (Coster et al., 2018; Zhang & Boyer, 2012).

Nursing research units are essential pillars to the preparation of nursing students at different levels of education and training and create conditions for advancing research processes. It is an excellent way to build solid lines of investigation that lead to the development of scientific knowledge within the discipline to obtain better health outcomes for the population and, consequently, influence educational processes, clinical practices, and health policies (Yanbing et al., 2021). On the other hand, nursing research units contribute to the development of other disciplinary areas. Interdisciplinarity and multidisciplinarity in healthcare sciences highlight the discipline of nursing within other health areas (Rodrigues, 2018; Yanbing et al., 2021; Zhang & Boyer, 2012). An emerging path is essential for the evolution of professions in the healthcare domain and the quality of care that the population needs and deserves. Therefore, research is a curricular component of crucial importance in the education of students at different levels of education.

Research promotes a wide range of results. Not only improves pedagogical processes, when investigating nursing pedagogy that leads to innovative learning and training practices but also when researching clinical nursing practices toward the promotion of innovative nursing care, increasingly focused on the real needs of the citizen (Yanbing et al., 2021; Zhang & Boyer, 2012).

4.1 How to bring the best evidence to the nursing teaching process?

All over the world, research units are dedicated to developing the nursing discipline, allowing for differentiating practices and products (National Research Council, 2011). The *National Institute of Nursing Research* is a great example of a nursing research unit in the United States that exists for over 35 years (Zhang & Boyer, 2012).

All this scientific development that has been conducted around nursing allows members of its professional body to be recognized as experts in their domain (Yanbing et al., 2021). Undoubtedly, nursing along the historical-scientific path has asserted itself as a science, playing a specific role within the multiprofessional context in the healthcare domain, largely due to the research that was and is instigated and conducted.

The development of nursing curricula in the research domain has been a driving force of change that should continue to foster the development of best practices, which simultaneously begin to trace new paths for conducting research (Coster et al., 2018).

For disciplinary knowledge to assert itself, curricula will need to be integrated with the best scientific methodologies for nursing research to reach an increasingly multidisciplinary matrix. The research will certainly be more collaborative across disciplines. Therefore curricula need to prepare students for these collaborative research processes.

Additionally, for the establishment of modern nursing, the practice of care or the health promotion processes of people/populations must be based on the best available evidence. For this to occur, robust and solid research processes are required to develop, improve, and expand knowledge (Coster et al., 2018).

Furthermore, nursing professionals must learn to anchor their actions on scientifically proven knowledge and seek the best available information based on the synthesis of the best available evidence, balanced with the preferences of the users of nursing care (Pearson et al., 2005). Only in this way will care be centered on the person's needs and resources and evidence-based, thereby reinforcing nursing and nurses' social value and professional specificity (Apóstolo, 2017; Wakibi et al., 2021; Yanbing et al., 2021).

In the same line of thought, another challenge for the pedagogical processes and therefore for nursing curricula is their capacity to be based on the best available knowledge, in addition to the need to be flexible enough to integrate the reinforcement and updating of practice and research skills (Wakibi et al., 2021). As this may be one of the best strategies to facilitate students' skill development at different learning and training levels, they will know how to use and critically contribute to the best evidence (Wakibi et al., 2021). Nursing must increase efficiency and effectiveness in the care it provides to the person, group, or community while preserving its character of compassion and humanity. Several world structures have promoted methodologies to support the central implementation of evidence synthesis in the international domain. In this context, and considering our discipline, the JBI is certainly a reference center for Nursing. JBI "develops and delivers unique evidence-based information, software, education and training designed to improve healthcare practice and health outcomes".[a] In addition to the vast methodologies already defined and made available in open access, this center is also increasingly focused on promoting methodologies to the transfer and implement science in clinical settings to complete the science cycle (Apóstolo, 2017; Aromataris & Munn, 2020; Porritt et al., 2020).

Therefore, evidence-informed care consists of doing the right things and ensuring that it is done well, attending to the user's preference and the nurse's clinical judgment. The best evidence emerges as an essential ethical component in the nursing profession, a social imperative valid for all human conduct.

[a] https://jbi.global/.

Several reasons justify the application of evidence-based nursing practice. They include: (1) favoring the incorporation of research results into professional practice; (2) promoting a more effective, efficient, and meaningful practice; (3) promoting the updating of knowledge based on the best levels of evidence; (4) optimizing resource administration; (5) fostering the continuous training of health professionals; and (6) promoting the encounter between professionals from different disciplines, degree of experience, and knowledge (Pearson et al., 2005; Wakibi et al., 2021).

When discussing the implementation of the best available scientific evidence into daily practice, the often-mentioned '60-30-10' digits emerge to portray a challenging scenario, specifically revealing that as much as 60% of all healthcare practice is indeed guided by evidence- or consensus-based guidelines. However, 30% is of low value due to research flaws and inconsistencies and 10% might actually be harmful (Braithwaite, Glasziou, & Westbrook, 2020). The so-called research waste is an issue particularly emphasized in recent years given the amount of evidence produced on a daily basis that poses many challenges when attempting to guide management and clinical decisions upon that evidence (Chalmers & Glasziou, 2009). On the one hand, there is the issue of poor research designs, conduct, and analysis (e.g., inadequate selection of research questions, insufficient attention to previous research results, weak research protocols) (Bleijenberg et al., 2018). This particular issue poses great challenges to both the newly graduate nurse and the Advanced Practice Nurse in terms of adequate processing of the quality of the evidence that reaches them and demands great awareness and confidence in research methods. On a clinical daily basis, the room for such evaluation processes is very reduced. Evidence summaries might be an adequate solution to start tackling this challenge (Siemieniuk & Guyatt, 2019). JBI is one of the most well known institutes worldwide embodying this challenge and producing evidence summaries and recommendations for best practices upon the best available evidence. As such, these institutions take upon them the role of gathering, filtering, and qualifying high-quality evidence so as to facilitate its implementation in daily practice.

On the other hand, researchers point to a mean wait time of about 17 years between innovation and its application in clinical practice (Balas & Boren, 2000), reinforcing the need to focus on implementation science as much as effectiveness research, in order to reduce the research waste. Implementation science has indeed received particular attention in the last decade. Several frameworks, models, and theories have been highlighted and connected to implementation work (Nilsen, 2015). The Getting Evidence into Practice (GRiP) approach is an example of a strategy to facilitate the implementation of best practices of particular utility to nurses pursuing the improvement of the quality of care (Munn et al., 2020).

Society needs nursing care underpinned by scientific evidence, not rituals or traditions, for all of the above. Advanced training in nursing, linked to the use of new technology, will facilitate these pedagogical processes to seek constant innovation and translation of scientific knowledge into practice.

5 Digital health in Advanced Practice Nursing

The use of information and communication technologies (ICT) in healthcare is recognized in terms of eHealth or Digital Health (WHO, 2023) and its adequate implementation has brought improvement to healthcare accessibility by reducing communication barriers related to geographical isolation or reduced resources (Melchiorre et al., 2018).

Along with the rapidly growing access of citizens and healthcare systems to ICT, a global need has also emerged for healthcare professionals and particularly nurses, to acquire digital and technological competences as prerequisites to carry out their professional roles (Konttila et al., 2019). Accordingly, advanced practice nurses are required to have developed nursing informatics competences, which makes nursing informatics a demanded field in advanced nursing curricula (Ahonen et al., 2016). Toward a nursing informatics competence framework, several key areas have been identified (Konttila et al., 2019). Nurses need to acquire knowledge of digital technology and digital skills required to deliver care of quality to patients, fundamentally developed upon social and communication skills as well as ethical considerations related to the eHealth realm. Once in the healthcare context, there is a need for collegial and organizational support in order to allow for the successful implementation of digital solutions in healthcare.

6 Innovative interaction strategies in nursing training

Competence development in nursing requires addressing several skills related to many dimensions, among which we emphasize (a) the clinical decision-making skills regarding nursing diagnosis, interventions, and outcomes (Melo, 2021); (b) the communication skills related to reporting skills on nursing consultation and therapeutic relation (Melo, 2021); and (c) the abilities to solve complex problems, based on scientific evidence and clinical intuition (Chien, 2019).

To facilitate nursing students' development of abilities related to the complex archetype of nursing approach to individuals, families, and communities, varied pedagogical strategies that address the dimensions mentioned above are demanded. The following text elicits examples of such pedagogical strategies that may be applied toward the facilitation of skill development in nursing education, specifically: (a) problem-based learning (PBL), (b) service-based learning (SBL), and (c) therapeutic interactions simulation and leadership skills promotion practices.

6.1 Problem-based learning

As a teaching method, PBL facilitates learning through reflection upon real-world situations, promoting critical thinking skills and the ability to focus on solutions rather than on problems. Case studies can present the problematic situation to students and work toward the solution, individually or in a group. While attending to and integrating the evidence, they are invited to search and identify as relevant (Duch, Groh, & Allen, 2001). These educational activities entail adopting specific approaches targeted at citizen or patient preferences and needs. Students are, therefore, invited to identify solutions for complex issues, such as caregiver role stress, community management, communication problems in the family, and structured issues, such as lack of knowledge or abilities related to specific roles (such as parental role, caregiver role, etc.).

6.2 Service-based learning

SBL combines theoretical approaches in the classroom with living experiences in real-world clinical settings (Rautio, 2012; Roehlkepartain, 2009). Such real-world settings might be community health services, where the proximity to the person's everyday context facilitates students' development of diagnostic and interventional skills at a person's private residency, schools, elder residencies, companies, prisons, etc. Patients may be invited to identify their resources and needs from a person-centered care perspective in the hospital setting. The opportunity to identify real-world healthcare settings gives students a sense of integrating learning with clinical practice using these methods. In nursing education, the SBL is a mandatory pedagogical strategy, particularly considering the clinical nature of nurses' endeavors and profession.

6.3 Simulation of therapeutic interactions and promotion of leadership skills practices

Roleplaying is an acknowledged methodology that facilitates the simulation of therapeutic interactions and promotes leadership skills practices, as it allows students to develop critical thinking and decision-making skills. Students may experience different interactions from a simulated scenario, fostering different solutions at the simulation labs (Dorri, Farahani, Maserat, & Haghani, 2019). By the experience of being in someone else's shoes (i.e., citizens, families, groups, or communities), where client and nurse (or other participants) twist their current role, a sense of empathy and humanization is promoted, in addition to clinical decision-making skills, derived from simulated real-life experiences.

In an increasingly complex and global world, introducing new techniques and tools in nursing education and bridging the gap between tradition and innovation allows for the consolidation of already tested and established results and the emergence of new evidence of innovative processes.

Intentional interaction training in the therapeutic relationship requires, for example, the development of consultation simulation moments. Such moments will occur in the clinical simulation rooms where students can observe, comment, and reflect on their observations. In parallel, they might as well experience the simulated scenario, self-analyze their performance, and discuss the reflections of colleagues and facilitators.

Different simulation contexts, created with different pedagogical purposes, will expose students to specific communication and clinical decision process challenges. Through roleplaying, such scenarios will promote the development of skills in conflict management, collaborative work, negotiation, and even clinical decision, derived from empirical knowledge and theoretical structures previously approached. Similarly, new learning needs might emerge, leading the student to seek knowledge and evidence to apply a posteriori.

The importance of simulation as a pedagogical innovation within nursing education has led to several developments. For example, in 2020, the Journal Clinical Simulation in Nursing was created by the International Nursing Association for Clinical Simulation and Learning - INACLS (https://www.inacsl.org/). INACLS further formed a committee to update the Healthcare Simulation Standards of Best Practice (Watts et al., 2021), where a stepwise approach is described to promote the implementation of simulation-based experiences (SBEs) (Table 1) (Watts et al., 2021:15).

The described steps must be applied in such a context that allows the training of simulation facilitators, the development of conditions to access the best evidence (including literature reviews and expert consultation), the preparation of spaces with a desk and/or chairs for training in therapeutic communication for instances, or more technological environments for simulating emergency situations or intensive care.

The richness of the nature of clinical decision-making processes in nursing is so vast and complex that the different methods of innovative interaction in nursing education

Table 1 Stepwise approach to the implementation of SBE.

Step	Actions
1	Co-design SBE with content- and simulation experts knowledgeable in best practices of simulation (including education, pedagogy and practice)
2	Perform needs' assessment to provide foundational evidence for a well-designed SBE
3	Construct measurable objectives that build upon the learner's existing knowledge
4	Build the simulation-based experience to align the modality with the objectives.
5	Design a scenario, case, or activity to provide the context for the simulation-based experience
6	Use various types of fidelity to create the required perception of realism
7	Plan a learner-centered facilitative approach driven by the objectives, learners' knowledge and level of experience, and the expected outcomes
8	Create a prebriefing plan that includes preparation materials and briefing to guide participant success in the simulation-based experience.
9	Create a debriefing or feedback session and/or a guided reflection exercise to follow the simulation-based experience.
10	Develop a plan for evaluation of the learner and of the simulation- based experience
11	Pilot- test simulation-based experiences before full implementation

Adapted from Watts, P.I., et al. (2021). Onward and upward: Introducing the healthcare simulation standards of best Practice™. *Clinical Simulation in Nursing, 5*, 14–21. https://doi.org/10.1016/j.ecns.2021.08.009.

must consider the nature of the skills to be developed in order to address specific inter-action processes with people who will be the target population of nurses' care, specifically:

- Intentional processes—require methods of developing clinical decision skills and intervening in knowledge, beliefs, values, attitudes, or adherence behaviors;
- Unintentional processes—require methods that involve clinical decision training and actions focused on the physiological aspects of people (such as wound care, intubation, and assessment of vital signs), as well as on emotions (for example, training in psycho-therapeutic actions);
- Processes of interaction with the environment—promoting pedagogical methods for culturally and politically sensitive care, conflict management, and leadership, among others related to interactive processes with different environments.

All of these aspects require nursing education to adapt to a context of major challenges associated with social and technological evolution, promoting innovative methodologies that enable the training of nurses with critical awareness, the ability to focus on solutions to complex problems, and a spirit of leadership and teamwork, in a constantly changing world, which demands dynamism, flexibility, and creativity from nursing facilitators and mentors.

7 Active learning sections

Critical thinking skills are fundamental to developing highly personalized and evidence-based nursing care (Hoke & Robbins, 2005). During nursing training, active learning methods are essential in acquiring these skills. Thus, and according to the purpose of this book, and specifically this chapter, we challenge the reader to interact by conducting the exercises presented below.

7.1 Suggested teaching assignments
7.1.1 Exercise 1
Considering the PcPF (McCormack & McCance, 2017) and the JBI systematic reviews' approach, develop a research problem based upon the theme of "Person-Centered Care" and a topic of your interest.

Accordingly, develop a search strategy that meets the purpose and type of secondary study the research problem aims to investigate (e.g., Population Concept Context—PCC, Population Intervention Comparator Outcome—PICO) and complete the suggested tables (Tables 2 and 3). The reader can adapt the table to a different systematic review approaches (e.g., PICO). Apply the search to MEDLINE (via PubMed) or adapt to a database of your choice.

Table 2 Example of a search strategy for a PCC-based problem.

	Description	MeSH https://www.ncbi.nlm. nih.gov/mesh	Keywords PubMed	Junction MESH+ Keywords
Population Concept Context Junction				

Source: Own elaboration.

Table 3 Example of a search strategy for a PICO-based problem.

	Description	MeSH https://www.ncbi.nlm. nih.gov/mesh	Keywords PubMed	Junction MESH+ Keywords
Population Intervention Comparator Outcome Junction				

Source: Own elaboration.

Choose an article that best suits your research problem and apply a checklist of a reporting guideline searched on the EQUATOR Network (https://www.equator-network.org/) and specific to the type of study of the article chosen.

7.1.2 Exercise 2

Search for a clinical nursing guideline or evidence summary (e.g., https://rnao.ca/bpg). According to the PcPF, your work context, and a specific population of your choice, adapt the chosen guideline.

Establish a working group to implement the guideline (part of step 1 of the JBI Implementation Project).

7.2 Recommended readings

Clinical Simulation in Nursing: https://www.inacsl.org/clinical-simulation-in-nursing-journal

JBI Evidence Synthesis Manual: https://jbi-global-wiki.refined.site/space/MANUAL

JBI Evidence Implementation Manual: https://jbi-global-wiki.refined.site/space/JHEI

7.3 Innovative tool

International Nursing Association for Clinical Simulation and Learning (INACLS): https://www.inacsl.org/

Caring for the Caregiver:

https://www.cabhi.com/blog/caring-for-the-caregiver-why-its-important-and-5-innovations-making-it-possible/

https://www.laboci.pt/en-gb#AboutUs

https://internationalcarers.org/carer-facts/global-carer-stats/

8 Synthesizing: From Advanced Pratice Nursing to nursing advanced training

In an evolving healthcare ecosystem, the concept of Advanced Practice Nursing extends beyond direct care to include leadership, mentorship, and the ability to incorporate evidence-based practices. The focus is on equipping nurses with the skills to address the diverse needs of individuals, their families, and the wider community. Advanced training in nursing serves as a conduit that enables nurses to transition from traditional roles to more specialized and advanced ones, supporting their capacity to manage complex health issues, make informed decisions, and lead healthcare teams effectively. At the heart of this transition is the person-centered curriculum, tailored to ensure the delivery of care that respects and responds to individual patient preferences, needs, and values. This curriculum fosters critical thinking, enhances interpersonal communication, and promotes an in-depth understanding of healthcare systems and policies. In doing so, it cultivates nurses who are competent, confident, and capable of delivering high-quality, person-centered care. Advanced Practice Nursing and nursing advanced training are inherently interconnected, each enriching and enhancing the other. Theoretical knowledge drawn from advanced training melds with practical insights gained from advanced practice nursing, facilitating a continuous cycle of learning, evolving, and improving in nursing care. The role of innovative pedagogical strategies in nursing advanced training is integral to this symbiotic relationship. Active learning methods, such as problem-based learning, service-based learning, and simulation of therapeutic interactions, present real-world scenarios that augment the learning process. These strategies foster critical thinking, problem-solving, and decision-making skills, thereby enabling nurses to adeptly manage intricate health situations. In the current healthcare environment, the digital health curriculum is becoming increasingly pivotal. The integration of digital technologies in nursing education enables access to up-to-date information, enhances learning through simulation and virtual reality, and fosters the development of digital competencies. This is crucial in today's healthcare milieu, where technology is rapidly evolving and playing a progressively significant role in healthcare delivery. Advanced Practice Nursing,

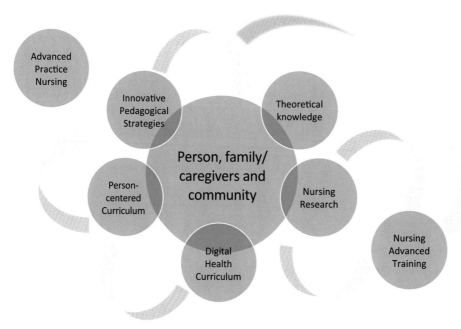

Fig. 1 From Advanced Practice Nursing to Nursing Advanced Training (Own elaboration).

nursing advanced training, and person-centered curriculum, bolstered by innovative pedagogical strategies and digital health curriculum, is essential in preparing nurses for the challenges of modern healthcare. With a strong grounding in theoretical knowledge and a focus on nursing research, nurses can enhance their skills, improve patient outcomes, and contribute to the ongoing development of healthcare systems. The future of healthcare hinges on well-prepared, competent, and dynamic nursing professionals, and this synthesis (Fig. 1) is a crucial stride toward realizing this vision.

9 Conclusions

In modern healthcare systems, healthcare professionals need to be agents of change and continuously integrate and balance the knowledge coming from the best available evidence with the knowledge of the person's values, preferences, and beliefs in a symmetrical partnership.

In this chapter we highlighted the challenges addressed in Advanced Practice Nursing for sustainable healthcare systems, based on the needs of the citizens, caregivers, and community and the impact of this on a person-centered curriculum. We also emphasized the value of research in nursing curricula and pedagogy based on the best evidence using innovative interaction strategies in nursing training: problem-based learning, service-based learning, simulation of therapeutic interactions, and promotion of leadership skills practices. We also presented active learning sections that can help educators and nursing students/nurses in their nursing advanced training.

We suggest continuous investment in this topic, carrying out research that shows the gains in the sense of Advanced Practice Nursing.

References

Ahonen, O., Kouri, P., Kinnunen, U. M., Junttila, K., Liljamo, P., Arifulla, D., & Saranto, K. (2016). The development process of eHealth strategy for nurses in Finland. In *Nursing informatics* (pp. 203–207).

Anagnostou, D. (2021). Nursing interventions improve preparedness, competence, reward and burden of family caregivers in end-of-life care at home. *Evidence Based Nursing*, *24*(1), 18. https://doi.org/10.1136/ebnurs-2019-103141.

Apóstolo, J. (2017). *Síntese da evidência no contexto da translação da ciência*. Coimbra, Portugal: Escola Superior de Enfermagem de Coimbra.

Aromataris, E., & Munn, Z. (Eds.). (2020). *JBI manual for evidence synthesis* JBI. https://doi.org/10.46658/JBIMES-20-01.

Balas, E. A., & Boren, S. A. (2000). Managing clinical knowledge for health care improvement. *Yearbook of Medical Informatics*, *09*(1), 65–70. https://doi.org/10.1055/s-0038-1637943.

Bartz, C. C. (2010). Conceptual explorations on person-centered medicine 2010: International Council of Nurses and person-centered care. International. *Journal of Integrated Care*, *10*(Suppl).

Bleijenberg, N., Janneke, M., Trappenburg, J. C., Ettema, R. G., Sino, C. G., Heim, N., … Schuurmans, M. J. (2018). Increasing value and reducing waste by optimizing the development of complex interventions: Enriching the development phase of the Medical Research Council (MRC) framework. *International Journal of Nursing Studies*, *79*, 86–93.

Braithwaite, J., Glasziou, P., & Westbrook, J. (2020). The three numbers you need to know about healthcare: The 60-30-10 challenge. *BMC Medicine*, *18*, 1–8.

Britten, N., Ekman, I., Naldemirci, Ö., Javinger, M., Hedman, H., & Wolf, A. (2020). Learning from Gothenburg model of person centred healthcare. *BMJ*, *370*.

Canadian Nurses Association. (2019). *Advanced practice nursing: A pan-Canadian framework*. Ottawa: Canadian Nurses Association.

Chalmers, I., & Glasziou, P. (2009). Avoidable waste in the production and reporting of research evidence. *The Lancet*, *374*(9683), 86–89.

Chesak, S. S., et al. (2021). A practice-based model to guide nursing science and improve the health and well-being of patients and caregivers. *Journal of Clinical Nursing*. https://doi.org/10.1111/jocn.15837.

Chien, L. Y. (2019). Evidence-based practice and nursing research. *The Journal of Nursing Research: JNR*, *27*(4), e29. . https://doi.org/10.1097/jnr.0000000000000346.

Clemmensen, T. H., et al. (2021). Informal carers' support needs when caring for a person with dementia – A scoping literature review. *Scandinavian Journal of Caring Sciences*, *35*(3), 685–700. https://doi.org/10.1111/scs.12898.

Coster, S., Watkins, M., & Norman, I. J. (2018). What is the impact of professional nursing on patients' outcomes globally? An overview of research evidence. *International Journal of Nursing Studies*, 76–83. https://doi.org/10.1016/j.ijnurstu.2017.10.009.

Dickson, C., Van Lieshout, F., Kmetec, S., McCormack, B., Skovdahl, K., Phelan, A., … Stiglic, G. (2020). Developing philosophical and pedagogical principles for a pan-European person-centred curriculum framework. *International Practice Development Journal*, *10*. https://doi.org/10.19043/ipdj.10Suppl2.004.

Dorri, S., Farahani, M. A., Maserat, E., & Haghani, H. (2019). Effect of role-playing on learning outcome of nursing students based on the Kirkpatrick evaluation model. *Journal of Education and Health Promotion*, *8*, 197. . https://doi.org/10.4103/jehp.jehp_138_19.

Duch, B. J., Groh, S. E., & Allen, D. E. (Eds.). (2001). *The power of problem-based learning*. Sterling, VA: Stylus.

Hamric, A. B., & Tracy, M. F. (2019). Chapter 3. A definition of advanced practice nursing. In M. F. Tracy, & E. T. O'Grady (Eds.), *Advanced practice nursing: An integrative approach* (pp. 61–79). St. Louis, MO: Elsevier.

Hoke, M. M., & Robbins, L. K. (2005). The impact of active learning on nursing students' clinical success. *Journal of Holistic Nursing, 23*(3), 348–355. https://doi.org/10.1177/0898010105277648.

International Council of Nurses (ICN). (2020). https://www.icn.ch/system/files/documents/2020-04/ICN_APN%20Report_EN_WEB.pdf.

Konttila, J., Siira, H., Kyngäs, H., Lahtinen, M., Elo, S., Kääriäinen, M., … Mikkonen, K. (2019). Healthcare professionals' competence in digitalisation: A systematic review. *Journal of Clinical Nursing, 28*(5–6), 745–761.

Langins, M., & Borgermans, L. (2015). *Strengthening a competent health workforce for the provision of coordinated/integrated health services.* Marmorvej.

Lloyd, H. M., Ekman, I., Rogers, H. L., Raposo, V., Melo, P., Marinkovic, V. D., … Britten, N. (2020). Supporting innovative person-centred care in financially constrained environments: The we care exploratory health laboratory evaluation strategy. *International Journal of Environmental Research and Public Health, 17*(9), 3050.

McCormack, B., & Dewing, J. (2019). International Community of Practice for person-centred practice: Position statement on person-centredness in health and social care. *International Practice Development Journal.*

McCormack, B., & McCance, T. (Eds.). (2017). *Person-centred practice in nursing and health care: Theory and practice* John Wiley & Sons.

Melchiorre, M. G., Lamura, G., Barbabella, F., & ICARE4EU Consortium. (2018). eHealth for people with multimorbidity: Results from the ICARE4EU project and insights from the "10 e's" by Gunther Eysenbach. *PLoS ONE, 13*(11), e0207292.

Melo, P. (2021). *Consultas de Enfermagem nos Cuidados de Saúde Primários: Guia de Decisão Clínica.* Lisboa: Lidel Editora.

Middleton, R., & Moroney, T. (2019). Using person-centred principles to inform curriculum. *International Practice Development Journal.*

Munn, Z., McArthur, A., Porritt, K., Lizarondo, L., Moola, S., & Lockwood, C. (2020). Evidence implementation projects using an evidence-based audit and feedback approach: the JBI implementation framework. In K. Porritt, A. McArthur, C. Lockwood, & Z. Munn (Eds.), *JBI handbook for evidence implementation* JBI. Available from: https://implementationmanual.jbi.global.

National Research Council. (2011). Research training in the biomedical. *Behavioral, and Clinical Research Sciences.*

Ng, R., & Indran, N. (2021). Societal perceptions of caregivers linked to culture across 20 countries: Evidence from a 10-billion-word database. *PLoS ONE, 16*(7), e0251161. https://doi.org/10.1371/journal.pone.0251161.

Nilsen, P. (2015). Making sense of implementation theories, models and frameworks. *Implementation Science, 10*(1), 1–13.

O'Donnell, D., McCormack, B., McCance, T., & McIlfatrick, S. (2020). A meta-synthesis of person-centredness in nursing curricula. *International Practice Development Journal, 10.* https://doi.org/10.19043/ipdj.10Suppl2.002.

Pan American Heath Organization and World Health Organization. (2003). *Core competencies for public health: A regional framework for Americans* (p. 40). Washington.

Parker, J. M., & Hill, M. N. (2017). A review of advanced practice nursing in the United States, Canada, Australia and Hong Kong Special Administrative Region (SAR), China. *International Journal of Nursing Sciences, 4*(2), 196–204.

Parmar, J., et al. (2021). Developing person-centred care competencies for the healthcare workforce to support family caregivers: Caregiver centred care. *Health & Social Care in the Community, 29*(5), 1327–1338. https://doi.org/10.1111/hsc.13173.

Parse, R. R. (2016). Where have all the nursing theories gone? [editorial]. *Nursing Science Quarterly, 29*(2), 101–102. https://doi.org/10.1177/0894318416636392.

Parse, R. R., Barrett, E., Bourgeois, M., Dee, V., Egan, E., Germain, C., ... Wolf, G. (2000). Nursing theory-guided practice: A definition. *Nursing Science Quarterly*, *13*(2), 177. https://doi.org/10.1177/08943180022107474.

Pearson, A., et al. (2005). The JBI model of evidence-based healthcare. *International Journal of Evidence-Based Healthcare*, *3*(8), 207–215. https://doi.org/10.1111/j.1479-6988.2005.00026.x.

Peterson, C. (2003). Health professions education: A bridge to quality. *Tar Heel Nurse*. https://doi.org/10.1111/j.1945-1474.2004.tb00473.x.

Phelan, A., McCormack, B., Dewing, J., Brown, D., Cardiff, S., Cook, N. F., ... McCance, T. (2020). Review of developments in person-centred healthcare. *International Practice Development Journal*, *10*. https://doi.org/10.19043/ipdj.10Suppl2.003.

Phelan, A., Rohde, D., Casey, M., Fealy, G., Felle, P., Lloyd, H., & O'Kelly, G. (2017). *Patient narrative project for person centred co-ordinated care*. Dublin: UCD/IPPOSI/HSE.

Porritt, K., et al. (Eds.). (2020). *JBI manual for evidence implementation* JBI. https://doi.org/10.46658/JBIMEI-20-01.

Rautio, A. (2012). Service-learning in the United States: Status of institutionalization. *Service Learning, General*, *139*. . https://digitalcommons.unomaha.edu/slceslgen/139.

Rodrigues, M. A. (2018). *Modelo Cross-cutting para gestão de atividades I & D e inovação: no caminho da moderna investigação em enfermagem* (pp. 141–154).

Roehlkepartain, E. C. (2009). Service-learning in community-based organizations: A practical guide to starting and sustaining high-quality programs. *Service Learning, General.*, *140*. . https://digitalcommons.unomaha.edu/slceslgen/140.

Ronquillo, C. E., Peltonen, L. M., Pruinelli, L., Chu, C. H., Bakken, S., Beduschi, A., ... Topaz, M. (2021). Artificial intelligence in nursing: Priorities and opportunities from an international invitational think-tank of the Nursing and Artificial Intelligence Leadership Collaborative. *Journal of Advanced Nursing*, *77*, 3707–3717. https://doi.org/10.1111/jan.14855.

Sabo, K., & Chin, E. (2021). Self-care needs and practices for the older adult caregiver: An integrative review. *Geriatric Nursing*, *42*(2), 570–581. https://doi.org/10.1016/j.gerinurse.2020.10.013.

Shin, J., et al. (2018). *Technology-mediated intervention to support Cancer caregiver*. ASCO EDUCATIONAL BOOK.

Siemieniuk, R., & Guyatt, G. (2019). What is GRADE. *BMJ J Best Practice*, *10*.

Wakibi, S., et al. (2021). Teaching evidence-based nursing practice: A systematic review and convergent qualitative synthesis. *Journal of Professional Nursing*, 135–148. https://doi.org/10.1016/j.profnurs.2020.06.005.

Watts, P. I., et al. (2021). Onward and upward: Introducing the healthcare simulation standards of best Practice[TM]. *Clinical Simulation in Nursing*, *5*, 14–21. https://doi.org/10.1016/j.ecns.2021.08.009.

WHO. (2023). *Global Observatory for eHealth*. https://www.who.int/observatories/global-observatory-for-ehealth.

World Health Organization (WHO). (2006). *Working together for health*.

Yanbing, S., et al. (2021). The state of nursing research from 2000 to 2019: A global analysis. *Journal of Advanced Nursing*, *77*(1), 162–175. https://doi.org/10.1111/jan.14564.

Zhang, Y., & Boyer, K. (2012). *National Institute of Nursing Research (NINR), Encyclopedia of global health*. https://doi.org/10.4135/9781412963855.n836.

CHAPTER 14

Next generation healthcare education and research: Utilizing the talent, skills, and competencies for value-based healthcare

Miltiadis D. Lytras[a] and Basim S. Alsaywid[b]
[a]College of Engineering, Effat University, Jeddah, Saudi Arabia
[b]Saudi National Institute of Health, Riyadh, Saudi Arabia

1 Healthcare education and research

The quest for digital healthcare or smart healthcare (Alsanea, 2012; Lytras, Papadopoulou, & Sarirete, 2020; Lytras, Sarirete, & Stasinopoulos, 2020a, 2020b) requires the utilization of diverse resources, including emerging technologies, best practices, and sustainable policies. In the era of artificial intelligence (AI) (Sairete et al., 2021), the digital transformation of work and education has to be seen as the physical evolution of the current state of healthcare delivery (Belliger & Krieger, 2018; Cresswell et al., 2013; Deloitte, 2018; Gopal, Suter-Crazzolara, Toldo, & Eberhardt, 2019; Haggerty, 2017). In this direction, technologies such as virtual reality and augmented reality (Alhalabi & Lytras, 2019; Al-Rasheed et al., 2022) or big data (Almasoud, Al-Khalifa, Al-salman, & Lytras, 2020) can be bold enablers. They are also demanding attention and resources for deployment in the education and training sectors (Kraus, Schiavone, Pluzhnikova, & Invernizzi, 2021; D.M. Lytras, Lytra, & Lytras, 2021; M.D. Lytras, Chui, & Visvizi, 2019).

In parallel, healthcare training institutions or organization that play a significant role in promoting health specialties and health educational and research skills deliver professional surveys on factors that affect the quality of healthcare education. For example:
- Alsaywid et al. (2020) elaborated on resident training in COVID-19 pandemic times with an integrated survey of the educational process, institutional support, anxiety, and depression by the Saudi Commission for Health Specialties (SCFHS).
- Hejazi et al. (2022) commented on the attitudes and perceptions of health leaders for the quality enhancement of the work force in Saudi Arabia.

Other areas of interest and questioning include concepts related to key performance indicators (KPIs), psychological factors including job satisfaction, anxiety, etc.:

Active Learning for Digital Transformation in Healthcare Education, Training and Research
https://doi.org/10.1016/B978-0-443-15248-1.00005-9

— Housawi et al. (2020a) published an evaluation of key performance indicators (KPIs) for sustainable postgraduate medical training; this was an opportunity for implementing an innovative approach to advance the quality of training programs at the SCFHS.
— Housawi et al. (2020b) also elaborated on a progressive model for quality benchmarks of trainee satisfaction in medical education toward the strategic enhancement of residency training programs at SCFHS.
— Housawi et al. (2021) also commented on a high-level strategy for implementing artificial intelligence at the SCFHS.

It seems that as Arafat, Aljohani, Abbasi, Hussain, and Lytras (2019) commented, there is a value connection among e-learning, web science, and cognitive computation that can support a new generation of learning services and learning analytics capable of improving the quality of education (Lytras, Serban, Ruiz, Ntanos, & Sarirete, 2022; Misseyanni, Marouli, Papadopoulou, Lytras, & Gastardo, 2016; Misseyanni, Papadopoulou, Marouli, & Lytras, 2018; Naeve, Yli-Luoma, Kravcik, & Lytras, 2008; Papadopoulou & Lytras, 2021; Papadopoulou, Lytras, Misseyanni, & Marouli, 2017; Sairete et al., 2021; Spruit & Lytras, 2018). Thus, the effectiveness and the preparedness of institutions to adopt technology-enhanced learning methods is a challenge beyond the COVID-19 pandemic (Alsaywid et al., 2021).

We live in times of change and fast development, especially in the domains of healthcare research and education. In such a demanding context, the provision of high-quality healthcare services has to be grounded in the excellent quality of healthcare education and research. While there is great discussion in the relevant literature for this phenomenon, in this concluding chapter we summarize some of our thoughts derived from our ongoing experiences and interactions with the relevant domain, both from an educational and a research perspective. In Fig. 1, we provide a graphical overview for our key proposition on the core determinants of Next Generation Healthcare Education and Research.

Research strategy: The research strategy is an integral part of active and transformative learning and research for all healthcare education institutes. The organizational-wide strategy should secure sufficient funding, resources, and policies for the utilization of research with significant social impact. For this purpose, the codesign of the strategy should include all the stakeholders of the institution, both internally and externally. Also, the researchers and faculty should be engaged on a transparent execution and assessment plan. The strategy should be oriented toward problem solving. Next-generation healthcare research should also look for multipliers of added value.

Education strategy: The education strategy jointly with the research strategy cultivates a joint footprint of next-generation healthcare education. With emphasis on the development of skills and competencies and utilizing talent, the educational strategy serves as the multiplier of value delivery. The main focus of the education strategy should be on the introduction of new multidisciplinary programs, the exploitation of technology-enhanced learning, robust professional development programs for faculty, and sufficient resources for educational strategy implementation.

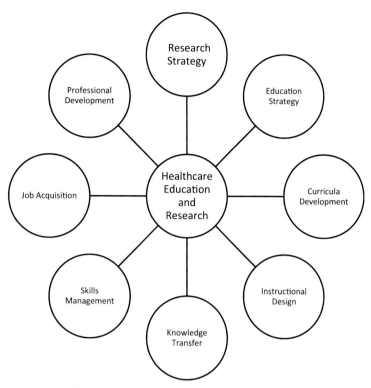

Fig. 1 The determinants of next-generation healthcare education and research.

Curricula development: New educational and training programs should be seen as a bold response to the next-generation active learning strategy in healthcare education. The development of curricula should be based on feasibility studies that will reflect the new requirements and the new needs in the healthcare industry. The design of training programs to reflect the new generation of jobs together with programs aiming to serve new professions due to the impact of Medical Technology on the classical skillset of healthcare specialties, is another bold direction for revisions of educational programs.

Instructional design: The new era of instructional design should be active and transformative as well as oriented toward learning. The fast delivery of the body of knowledge in the healthcare specialties with the utilization of open knowledge repositories and new tools should also be supported with fresh ideas and methods on instructional design. The dynamic exploration of health-related scientific knowledge, the dynamic composition of personalized learning paths and spaces, and the enhanced engagement of learners must be bold priorities. There is also a need for a significant increase in professionals aiming to serve a highly demanding instructional design process within higher education healthcare institutions.

Knowledge transfer: The establishment of knowledge dissemination mechanisms within the healthcare education institutions is another critical aspect for effective next-generation healthcare education and research. The design and implementation of knowledge management platforms aiming to codify and transfer best practices, the body of knowledge, and the latest developments and innovations in all areas of healthcare must be seen as a bold milestone toward the implementation of an active learning strategy.

Skills management: The detailed job profiling of all the healthcare specialties, and the systematic mapping of skills and competencies updated with a new set of items related to Medical Technology and digital health will serve as the required bridge between education and industry. The industry needs for highly skilled healthcare practitioners (HCP) will set new challenges for the redesign of skillsets and the promotion of a competency and skills-based education. From an educational point of view, this will also lead to a significant update of the agenda for educational strategy for a strategic alignment of education and research strategy to the industry priorities and the job market.

Job acquisition: The outlook of healthcare education and research to the healthcare industry and market will develop strong bonds of educational programs with job acquisition. The healthcare higher education institutions will have to support the facilitation of the job-seeking process with internal structures and processes. Well-educated healthcare practitioners should be exposed to job market opportunities from the early years of their educational programs.

Professional development: The professional development strategy should also be integrated in the active and transformative learning approach in healthcare education institutions. The life-long learning provision of timely and integrated professional development programs should be managed as a strategic stream of top-quality training programs with emphasis on building skills and competencies.

2 Talent, skills, and competencies management for value-based healthcare

One of the greatest challenges for next-generation healthcare education and research is the management of talent, skills, and competencies. In our effort to summarize our thoughts, we designed Fig. 2, which synthesizes our ideas on this matter. The purpose of this section is not to be exhaustive but to serve the scientific debate on the effective utilization of talent, skills, and competence-based education.

2.1 Talent management

For many years, the concept of talent management in healthcare education has been ignored systematically. The monolithic educational approaches and uniform training programs have rarely focused on issues related to the identification and enhancement

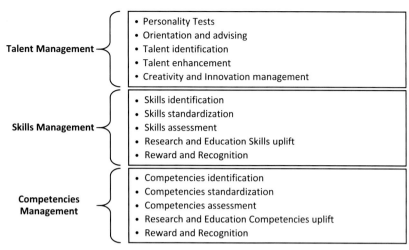

Fig. 2 A proposed taxonomy of items to be considered for talent, skills, and competencies management in the context of next-generation healthcare education and research.

of talent management. In an ongoing educational process with the delivery of static content and courses with a monodimensional approach, there is a need for introducing active and transformative learning practices that will allow the utilization of student and trainee talent. A few ideas on an organized way to respond affectively to the challenge of talent management are provided in the next paragraphs.

Personality tests can be a useful tool and practice for the identification and exploitation of personality traits that can be explored in unique and customized learning paths and programs. In this direction, the establishment within higher education of healthcare education institutions of capabilities for the parametrization of learning content delivery and exploration of content should be a priority. Furthermore, the systematic profiling and the development of systematic portfolios of student knowledge, talents, skills, and competencies can also be considered. The newly arrived idea of people's or students' analytics should be adopted to address the need to establish sophisticated talent and skills management systems in academia and research. The same consideration should also be applied to research talent.

Orientation and advising is an institutional process that must be supported with new policies, services, and community services. The early allocation of talented students to specialties, especially the medical ones, can be seen as an opportunity and not as a threat for the educational system. To this direction, new technology-enhanced learning services, repositories of lessons learned, stories, and sophisticated AI platforms for talent identification and enhancement can be introduced.

Talent identification: The talent identification process has to be established in healthcare education institutions. To our perception, talent is a rare resource that must be identified. Systems related to personality tests, identification of patterns over grading systems, and the exposure of young students to problem-solving scenarios are indicative ideas. Furthermore the next generation of healthcare with its bold technological components allows the identification of talent in new innovative ideas for services, healthcare startups, and Medical Technology systems.

Talent enhancement: A new generation of technology-enhanced learning services together with brand new ideas on the way that talent is enhanced can offer significant value to this area. The enhancement of the talent of young healthcare students or professionals should be seen as a continuous active and transformative learning development process. Rapid expertise transfer from the best professionals, the dissemination of timely best industrial practices related to healthcare, and the enrichment of talent with contributions from other domains such as AI are a few suggested actions and initiatives in this direction.

Creativity and innovation management: The next generation of healthcare will require new jobs, new services, and new value propositions defined by new needs and challenges. In this context, the idea of creativity and innovation management is aiming the applied aspect of the utilization of healthcare knowledge and practice at real-world problems. The promotion of creative thinking through active and transformative learning in healthcare can be formulated with different unique processes and services such as innovation competitions in healthcare, robust research-evidence innovation, and strong capacity building with the use of cloud services or crowdsourcing. A new perception for healthcare specialties must also be explored though new systems and services that exploit the collective intelligence of all the experts in the healthcare domain.

2.2 Skills management

The educational curricula for the healthcare specialties need to reassess and reconsider the skillsets that can effectively and jointly support the related job profiles. In this direction, the systematic identification, standardization, assessment of skills, and uplift of research and educational skills should be promoted and recognized with substantial rewards.

The following items are indicative in terms of skills management:
- Skills identification.
- Skills standardization.
- Skills assessment.
- Research and education skills uplift.
- Reward and recognition.

Beyond this list, objectives related to the acquisition of skills through education and training together with efforts at establishing systems for the dynamic allocation and optimization of skills on demand can support a new era of skills management in healthcare education.

- *Skills identification*: Skills identification can be multifold, from one side to provide a solid structure of items under skills categories and from the other side to assess the current skillsets of trainees and students. In this case, the educational and research effort can be seen as a sophisticated way of skills elevation.

- *Skills standardization*: The challenge of standardization related to skills is not a new practice. For a long time, areas related to human resource management and research and development projects have been targeted to the development of detailed lists of skills and competencies under different categories of core or soft skills. Also, in recent years, the development of analytics research along with the promotion of people's analytics have provided interesting insights into the way that skills can be standardized. The obvious benefit of this process is the development of smart services aiming to assess automatically or with the use of machine learning and AI services the skills management.

- *Skills assessment*: In the center of a competence and skills-based education is the systematic provision of learning interactions for the elevation of skills. Thus, the assessment of skills with formative or other types of assessment together with certification of skills becomes a priority. In this direction, the skills assessment can also be facilitated by services and active and transformative learning scenarios to measure soft skills and other capabilities of students and trainees.

- *Research and education skills uplift*: The next generation of healthcare education and research has to promote strategies, initiatives, and monitoring plans for the uplift and elevation of research and educations skills. This effort will require a clear understanding of the current situation and a vision of the new era. Having in mind that in the next years new skills will be developed related to the deployment of new technologies in healthcare practice and research, there is a clear elevation path related to skills and competencies related to the Medical Technology adoption. The other clear path is related to new creative ways of healthcare provision with an emphasis on innovative services and healthcare industries that currently may not exist. For example, consider the healthcare data lakes and the need to use sophisticated AI and machine learning methods to get value out of them.

- *Reward and recognition*: The efficient support of a robust skills and competencies management strategy in the context of next-generation healthcare education has to be based on a well-defined and communicated rewards and recognition policy. New practices and methods for recognition and reward related to skills assessment and certification can potentially lead to a healthy competition among healthcare practitioners for achievements. New technology-enabled services can also be used for certifying the impact of formal or informal professional development initiatives on student or professional profiles and skillsets. Consider professional social networks, where the skills and competencies of all students and healthcare practitioners are used for talent management, rewards, and professional recognition. The management of diverse structured and unstructured data elements would be a bold requirement toward this direction.

2.3 Competencies management

In close relation to the previous discussion on talents and skills, the analysis of the challenges related to the competencies management in the context of next-generation healthcare education and research can be organized around the following items:

- Competencies identification.
- Competencies standardization.
- Competencies assessment.
- Research and education competencies uplift.
- Reward and recognition.

3 Conclusions

Envisioning next-generation healthcare education and research is an art. In fact, it is about bringing into action ideas and creative thinking for the new value-driven healthcare proposition that directly affects the quality of life and the well-being of all of us. In this concluding chapter, we emphasized only the human aspects and mostly the core role of talents, skills, and competencies.

In this concluding chapter, we tried to communicate only a few of the ideas for the next generation Healthcare Education and Research. Our ongoing research and international collaborations are promoting further additional aspects of the phenomenon. Our ultimate objective is to promote the scientific dialogue for a next generation value-based healthcare education.

References

Alhalabi, W., & Lytras, M. D. (2019). Editorial for special issue on virtual reality and augmented reality. In *Vol. 23. Virtual reality* (pp. 215–216). London: Springer.

Almasoud, A., Al-Khalifa, H., Al-salman, A., & Lytras, M. (2020). A framework for enhancing big data integration in biological domain using distributed processing. *Applied Sciences, 10*(20), 7092.

Al-Rasheed, A., Alabdulkreem, E., Alduailij, M., Alduailij, M., Alhalabi, W., Alharbi, S., et al. (2022). Virtual reality in the treatment of patients with overweight and obesity: A systematic review. *Sustainability, 14*(6), 3324.

Alsanea, N. (2012). The future of health care delivery and the experience of a tertiary care center in Saudi Arabia. *Annals of Saudi Medicine, 32*(2), 117–120.

Alsaywid, B., Housawi, A., Lytras, M., Halabi, H., Abuzenada, M., Alhaidar, S. A., et al. (2020). Residents' training in COVID-19 pandemic times: An integrated survey of educational process, institutional support, anxiety and depression by the Saudi Commission for Health Specialties (SCFHS). *Sustainability, 12*(24), 10530.

Alsaywid, B., Lytras, M. D., Abuzenada, M., Lytra, H., Sultan, L., Badawoud, H., et al. (2021). Effectiveness and preparedness of institutions' E-learning methods during the COVID-19 pandemic for residents' medical training in Saudi Arabia: A pilot study. *Frontiers in Public Health, 9*, 707833.

Arafat, S., Aljohani, N., Abbasi, R., Hussain, A., & Lytras, M. (2019). Connections between e-learning, web science, cognitive computation and social sensing, and their relevance to learning analytics: A preliminary study. *Computers in Human Behavior, 92*, 478–486.

Belliger, A., & Krieger, D. J. (2018). The digital transformation of healthcare. In *Knowledge management in digital change* (pp. 311–326). Cham: Springer.

Cresswell, K., Coleman, J., Slee, A., Williams, R., Sheikh, A., & ePrescribing Programme Team. (2013). Investigating and learning lessons from early experiences of implementing ePrescribing systems into NHS hospitals: A questionnaire study. *PLoS ONE, 8*(1), e53369.

Deloitte. (2018). *The 2018 global health care outlook: The evolution of smart health care.* Deloitte.

Gopal, G., Suter-Crazzolara, C., Toldo, L., & Eberhardt, W. (2019). Digital transformation in healthcare– architectures of present and future information technologies. *Clinical Chemistry and Laboratory Medicine (CCLM), 57*(3), 328–335.

Haggerty, E. (2017). Healthcare and digital transformation. *Network Security, 2017*(8), 7–11.

Hejazi, M. M., Al-Rubaki, S. S., Bawajeeh, O. M., Nakshabandi, Z., Alsaywid, B., Almutairi, E. M., et al. (2022). Attitudes and perceptions of health leaders for the quality enhancement of workforce in Saudi Arabia. *Healthcare, 10*(5), 891.

Housawi, A., Al Amoudi, A., Alsaywid, B., Lytras, M., bin Moreba, Y. H., Abuznadah, W., et al. (2020a). Evaluation of key performance indicators (KPIs) for sustainable postgraduate medical training: An opportunity for implementing an innovative approach to advance the quality of training programs at the Saudi Commission for Health Specialties (SCFHS). *Sustainability, 12*(19), 8030.

Housawi, A., Al Amoudi, A., Alsaywid, B., Lytras, M., bin Moreba, Y. H., Abuznadah, W., et al. (2020b). A progressive model for quality benchmarks of trainees' satisfaction in medical education: Towards strategic enhancement of residency training programs at Saudi Commission for Health Specialties (SCFHS). *Sustainability, 12*(23), 10186.

Housawi, A., Alsaywid, B., Lytras, M. D., Apostolaki, A., Tolah, A. W., Abuzenada, M., et al. (2021). High-level strategy for implementing artificial intelligence at the Saudi Commission for Health Specialties. In *Artificial intelligence and big data analytics for smart healthcare* (pp. 11–23). Academic Press.

Kraus, S., Schiavone, F., Pluzhnikova, A., & Invernizzi, A. C. (2021). Digital transformation in healthcare: Analyzing the current state of research. *Journal of Business Research, 123*, 557–567.

Lytras, M. D., Chui, K. T., & Visvizi, A. (2019). Data analytics in smart healthcare: The recent developments and beyond. *Applied Sciences, 9*(14), 2812.

Lytras, D. M., Lytra, H., & Lytras, M. D. (2021). Healthcare in the times of artificial intelligence: Setting a value-based context. In *Artificial intelligence and big data analytics for smart healthcare* (pp. 1–9). Academic Press.

Lytras, M. D., Papadopoulou, P., & Sarirete, A. (2020). *Smart healthcare: Emerging technologies, best practices, and sustainable policies.* Elsevier. https://doi.org/10.1016/b978-0-12-819043-2.00001-0.

Lytras, M. D., Sarircte, A., & Stasinopoulos, V. (2020a). *Policy implications for smart healthcare: The international collaboration dimension.* Elsevier. https://doi.org/10.1016/b978-0-12-819043-2.00017-4.

Lytras, M. D., Sarirete, A., & Stasinopoulos, V. (2020b). Policy implications for smart healthcare: The international collaboration dimension. In *Innovation in health informatics* (pp. 395–402). Academic Press.

Lytras, M. D., Serban, A. C., Ruiz, M. J. T., Ntanos, S., & Sarirete, A. (2022). Translating knowledge into innovation capability: An exploratory study investigating the perceptions on distance learning in higher education during the COVID-19 pandemic—The case of Mexico. *Journal of Innovation & Knowledge.* https://doi.org/10.1016/j.jik.2022.100258.

Misseyanni, A., Marouli, C., Papadopoulou, P., Lytras, M., & Gastardo, M. T. (2016). Stories of active learning in STEM: Lessons for STEM education. In *Proceedings of the international conference the future of education* (pp. 232–236).

Misseyanni, A., Papadopoulou, P., Marouli, C., & Lytras, M. D. (2018). *Active learning strategies in higher education.* Emerald Publishing Limited.

Naeve, A., Yli-Luoma, P., Kravcik, M., & Lytras, M. D. (2008). A modelling approach to study learning processes with a focus on knowledge creation. *International Journal of Technology Enhanced Learning, 1*(1–2), 1–34.

Papadopoulou, P., & Lytras, M. D. (2021). Empowering the one health approach and health resilience with digital technologies across OECD countries: The case of COVID-19 pandemic. In *Artificial intelligence and big data analytics for smart healthcare* (pp. 225–241). Academic Press.

Papadopoulou, P., Lytras, M., Misseyanni, A., & Marouli, C. (2017). Revisiting evaluation and assessment in STEM education: A multidimensional model of student active engagement. In *EDULEARN17 proceedings* (pp. 8025–8033).

Sairete, A., Balfagih, Z., Brahimi, T., Amin Mousa, M. E., Lytras, M., & Visvizi, A. (2021). Editorial—Artificial intelligence: Towards digital transformation of life, work, and education. *Procedia Computer Science*. https://doi.org/10.1016/j.procs.2021.11.001.

Spruit, M., & Lytras, M. (2018). Applied data science in patient-centric healthcare: Adaptive analytic systems for empowering physicians and patients. *Telematics and Informatics*. https://doi.org/10.1016/j.tele.2018.04.002.

Index

Note: Page numbers followed by *f* indicate figures and *t* indicate tables.